MARINE CLIMATE, WEATHER AND FISHERIES

MARINE CLIMATE, WEATHER AND FISHERIES

The Effects of Weather and Climatic Changes On Fisheries and Ocean Resources

TAIVO LAEVASTU

Halsted Press: an Imprint of JOHN WILEY & SONS, INC.
NEW YORK TORONTO

Published in N. America by Halsted Press,
an Imprint of John Wiley & Sons, Inc.,
605 Third Avenue, New York, NY 10158-0012.

Library of Congress Cataloging-in-Publication Data
is available.

ISBN 0-470-22049 X

First published in Great Britain in 1993 by Fishing News Books
an imprint of Blackwell Scientific Publications
Osney Mead, Oxford OX2 0EL

Printed in Great Britain

Contents

Preface

The objectives of this book

Climatic change and its possible effects on life on the earth has received considerable worldwide attention recently, under headlines such as global warming and global change. It is necessary in consequence to examine what the effects of climatic changes and fluctuations are on marine life, and especially on marine fishery resources.

The marine climate is the integrated mean of the marine weather. However, the immediate effects of the environment and its changes on fish are transmitted by the action of weather elements and processes on the ocean environment. Therefore it is necessary to evaluate and understand the interactions of sea and air and the effects of weather on the ocean in different space and time scales. Practical fishermen need to know and understand the apparent relations between marine weather and the availability of fish and fluctuations in the abundance of fishery resources. Some of this knowledge has been used by fishermen throughout the ages to improve their catch. Understanding the short term effects of weather on fish will also permit us to understand the possible effects of climatic fluctuations on fish.

The catches of given species of fish have, in the past, often been assumed to represent the trend in abundance of the species and used as such in various correlation analyses to hypothesize about the effect of fishing and of climatic changes on stocks. However, the catches often reflect the trend in exploitation and are affected by economics, technical developments and fisheries regulations. Furthermore, the apparent collapses of fish stocks are affected not only by fishing and the ocean climate but by many other processes, such as pollution and other human activities, as well as natural processes within the fish ecosystem such as diseases and changing predation conditions. In order to evaluate the effects of climate on fish, we must attempt to separate the natural changes from those caused by man and reexamine past data and conclusions.

The objectives of this book fall into four categories: first, to give selected basic background on marine weather and climate and to describe how the surface weather affects the ocean environmental conditions and processes in the surface layers of the ocean; second, to summarize some of the effects of weather and ocean climate changes on fish stocks and on fish availability, and the use of this knowledge in practical fishing; third, to review and summarize existing knowledge on the possible effects of climatic fluctuations and man-induced effects on fish stocks; and fourth, to review some major historical changes in fisheries.

The basic purposes of the book are: to explain the fundamentals of the effects of surface weather on the surface layers of the ocean; to summarize briefly some practical knowledge of the effects of weather on fish behaviour and availability; and to examine critically the possible effects of climatic fluctuations on marine fishery resources. Thus the book is directed to practical fishermen, fisheries scientists, and especially as a background review to scientists and research administrators studying climatic change and its effects.

The sources and nature of information

Over the last century numerous correlations between landings of given species and environmental factors have been reported, and the variations in catches have often been attributed to climatic changes. However, many of the correlations found between landings and changes of some climatic elements have failed on further analysis to show that change in fish stock was caused by climatic change. Coincidental changes may cause empirical relationships to give incorrect indications. Many of the major classical interactions between climate and fisheries have been summarized in an excellent book by Cushing (1982a). He also pointed out that simple correlations tend to break down when more information becomes available.

In writing this book an attempt is made to understand and interpret the environment–fish relationships and to report the results of previous studies which can be explained on the basis of functional relationships (i.e. on cause–effect principles). As many of the stock abundance and/or availability fluctuations are local by nature, no attempt is made to analyse them in detail in this book. Rather, it attempts to discover whether climatic changes might be the cause of fluctuations and by what mechanism.

Some of the possible mechanisms for the effects of climate on fish stocks can be arrived at on theoretical grounds, but definite proof is difficult to obtain. For example, Corten (1990) found that although some long-term changes in pelagic fish stocks in the North Sea could be theoretically explained by assuming changes of flow of Atlantic water into the North Sea, there is no physical evidence that such changes in the inflow of Atlantic water occurred.

Cushing (1982a) described historic sporadic appearances and disappearances of fishes in given locations. The causes of most of these are uncertain in the sense that we cannot ascertain to what extent they resulted from changes in stock size or from changes in annual and life-cycle migration routes caused by a number of possible environmental factors.

Very few scientific studies have been conducted on the effects of weather on the availability and abundance of fish. Knowledge of the relationship between weather and ocean conditions and fish behaviour and availability is gained and used daily by fishermen to locate fishable aggregations and in various tactical fishing decisions to increase the catch per effort spent.

Fishing skippers usually regard this knowledge as privileged information, and the basis of their skills in knowing how to make use of it. This knowledge is also very diverse and is dependent on location and type of fishery, especially in coastal fisheries. Only a general review of this subject can, therefore, be presented in this book.

It is often claimed that if the effects of surface weather on ocean environment and on fish stocks were known, one could predict the occurrence of fish and their availability to gear, from the weather and ocean predictions. We could use this hypothesis only for short term prediction of fish availability. Longer term predictions of over a week are not possible, because weather and other environmental forecasts over 1 week have no predictive value.

A large number of published reports describe possible relationships between climate and fisheries in the past. Some of this knowledge has become general and hypothetical, and many reported results need further scrutiny. It is quite common to assume that climate might have affected the fluctuations of landings, when the possible effects of fishing do not explain the observed fluctuations. However, no full documentation of the climatic change or explanation of how this change might have affected the stock is presented in the majority of the reports. As several factors simultaneously affect the landings as well as fluctuations in stock abundance, a quantitative separation of the effects of the factors is, in most cases, very difficult if not impossible. No attempt has been made in this book to reanalyse reported data, but attention is called to possible reconsideration of reported results in many cases.

This book is organized as follows. Chapter 1 describes how the weather and climate affect fish, fishing and landings. The ways in which biological processes within the fish ecosystem influence stock fluctuations are summarized, and the climatic and hydroclime elements which might affect fish and fishing and the nature of their changes are briefly reviewed. The main purpose of this section is to outline the chief processes and elements of weather, climate and fishery interaction.

Chapter 2 reviews the essential marine surface weather processes and how these change the environmental conditions in the surface layers of the sea. Chapter 3 describes the relationships between marine climate and hydroclime elements and processes, and the behaviour and fluctuations in abundance of fish stocks.

Chapter 4 summarizes the practical aspects of the effects of weather on fish finding and fishing operations. Chapter 5 examines how the effects of fishing on stocks and on landings can be separated from the effects of climatic fluctuation, and reviews the past fluctuations of neritic stocks and their possible causes.

Chapter 6 describes the effects of eutrophication and pollution, including oil pollution, on fish stocks and fisheries. The final Chapter summarizes major historical changes in fisheries, evaluates briefly the possible causes of these changes and summarizes the past and future aspects of possible climate changes and fisheries interactions.

Acknowledgements

The author expresses his thanks to Mrs Marge Gregory for typing the manuscript and to Mr Bernard Goiney for preparation of figures. Furthermore, the author thanks his former colleague Dr Felix Favorite for valuable discussions and advice, and Mr Richard Miles of Fishing News Books for encouraging him to write this book, and Dr James Mason who had many good suggestions for clarification of the interwoven subjects in this book.

Chapter 1
Modes of Influence of Climate and Weather on Fisheries and Their Resources

The evaluation of the possible effects of weather and climate changes on fishery resources and fishing requires a prior knowledge of the possible modes and mechanisms of their influence on fish. This chapter briefly reviews these mechanisms, and evaluates in general terms their possible effects, i.e. how weather and climatic changes affect fishing and marine resources. Thus it is a qualitative search for the possible causes of changes in catches and in resource abundance and distribution. Many of these causative factors act concomitantly (e.g. changes of fishing effort and stock size) and many are gradual (e.g. improvement of gear). Consequently a quantitative separation of the effects of individual forces and factors is difficult and also variable in space and time.

First a brief review is given on how different weather elements can affect fishing and fish stocks, fish aggregation and availability. Thereafter the numerous factors which affect catches and landings are reviewed, as the latter are often used in the investigations of the effects of climatic changes on fisheries. Many cases are known in the past where stock of one species declined while the stock of another species increased. The causes of this change of dominance are often obscure, but could be ascribed to climatic influence; this phenomenon of species replacement is, therefore, also briefly reviewed.

The changes of stock size and/or availability to a fishery for a given species in a given region can be affected by many other factors besides weather and climatic fluctuations, and these are considered in this section in a general manner. The nature of changes of essential climatic elements, their space and time scales and nature of their effects on fisheries are also briefly reviewed.

Further review and evaluation of factors affecting fisheries and fish stocks includes the consideration of man-made effects of pollution and eutrophication, technical improvements of fishing vessels and gear and changes of market demands and prices. These last mentioned effects of diverse man-dependent factors on landings should be quantified and, whenever possible, separated from the effects of weather and climatic fluctuations on stocks.

1.1 Effects of weather elements on fish and fishing

The surface weather elements influence fish through their effect on the ocean, affecting the availability of fish to gear and aggregation of fish in different depths and areas. Most fishing occurs when fish are sufficiently aggregated, and the fishermen use their knowledge of the dependence of fish aggregation and availability on weather to improve their catches. This knowledge is usually based on their past observations and experiences, and fisheries research has dealt with this subject hardly at all.

Fisheries are more affected by weather than any other category of food production. Summaries of the effects of weather on fisheries have been given by Laevastu (1961) and Waterman & Cutting (1960). Weather affects the safety and comfort of fishermen. Modern larger fishing vessels can operate in winds up to force 8, but smaller vessels have to cease fishing at a lower wind force, usually force 6. On offshore grounds the vessels have to 'dodge' the weather and ride out the gales, whereas in coastal fisheries they search for safety in ports. Consequently landings of fish can be a function of frequency of storms, which vary seasonally and from year to year.

Storms affect the safety of fishing vessels via their effects on ocean surface, i.e. the waves. For example, dangerous situations for ships arise when the wave length approaches the ship's length, and especially if the ship is in a following sea. Icing of the superstructure in high latitudes is another dangerous condition which has caused the loss of many fishing vessels. Another weather element affecting fishing and safety of navigation is fog, especially in coastal areas. In subtropics and tropics high air temperatures affect the spoilage of fish, especially if there is lack of refrigeration or if delay in stowage occurs. Most coastal countries provide fishing and shipping with weather services, which are used in tactical decisions at the discretion of the skippers.

Weather also affects the fish behaviour, their aggregation, dispersal and migrations (vertical and horizontal) through its effects on the ocean. Wind, which causes turbulence and mixing in the surface layers of the oceans, influences the vertical movement of fish shoals, usually sending them into deeper layers below the oscillatory wave movement during heavy winds. Light, which is dependent on cloud cover among other factors, might also affect the distribution of fish with depth, especially species undertaking diurnal vertical migrations. Wind induced currents transport fish and fish eggs and larvae and might affect fish migrations. For example, brood strength of Georges Bank haddock is assumed to be predictable from the occurrence of offshore winds during the pelagic season of eggs and larvae (Chase 1955). Similarly Koslow *et al.* (1987) found that year-class success of cod and haddock in the NW Atlantic is somewhat associated with offshore winds. Whether these reported relations between wind and recruitment have been successfully used for prediction is not known.

Most variable weather elements are related to each other by functional processes. Similarly most variable processes and properties of surface layers of the ocean are related to weather and to each other. There is also a feedback from the ocean to the atmosphere (e.g. heat and water vapour) and anomalies in this feedback have been considered by some meteorologists to be the primary generator of short term climatic anomalies. The ocean is also a good integrator of the variable effects of weather, as the changes in the ocean are considerably slower than the corresponding changes in the near-surface layers of the atmosphere.

The more pronounced effects of weather on the ocean are those caused by surface wind, i.e. waves, mixing by waves within depth and creation of surface mixed layer thickness, and currents (wind component). Heating and cooling of surface layers of the ocean, as well as heat and moisture exchange between the sea and the air, are also greatly influenced by wind. The biology of plankton and its production are greatly affected by weather caused processes and changes in surface layers. A review of the marine environment as a determinant of fisheries has been given by Laevastu & Favorite (1988) and the effects of environmental factors on fish have been described by Laevastu & Hayes (1981). Further details on the effect of weather on ocean are given in Chapter 2, and Chapter 4 describes the effects of weather on fish availability and on fishing operations.

1.2 The nature of long-term changes in fisheries

Changes in catches and landings

The data (statistics) of catches and landings have been in most cases in the past used as an index of stock abundance in the studies of effects of climatic fluctuations on stocks. However, catches and landings are affected by many factors other than climatic changes and/or abundance and availability of fish to gear. In order to evaluate the effects of climatic changes on fisheries using catch statistics, all factors affecting catches must be recognized.

Hempel (1978a) and Holden (1978) summarized the fish catches from the North Sea and reviewed various attempts to explain their fluctuations by natural and man-made factors. They were, however, not always able to separate and quantify the effects of fishing from the effects of other factors on the fluctuations of stocks. Factors affecting landings are listed below in two categories, technical and economic factors and biological factors:

(1) Catches can change owing to alterations in fishing intensity (effort) (e.g. increases of fishing fleets).

(2) Economic factors (market demands and prices) can cause the fishing effort to alter and/or shift from one species to another.

(3) Changes of fishing vessel types and their propulsion can affect catches and cause shifts from one species to another.
(4) Likewise change of principal gear or introduction of new gear can influence catches and landings (e.g. the introduction of the power block had a major impact on the introduction of large, effective purse seines).
(5) Switching principal fishing grounds or discovering new ones can cause changes in species composition in catches as well as total catches.

There are biological factors in fish stocks which affect catches. These might be affected by climatic changes, but fluctuations in catches are usually caused by a combination of factors, including intensity of fishing:

(6) Shifts of population centres of species, and thus of the availability of fish, have been known for a long time, some of which cannot be fully explained.
(7) Fluctuation of recruitment occurs in nearly all stocks, the causes of which can be diverse but are usually factors and processes in the fish ecosystem itself, such as changes in predation on juveniles (in turn caused by many factors).
(8) There are many other factors affecting the fishing as well as stocks, including growth rate changes, local pollution, and lately a multitude of 'management actions'.

The reportings of landings, which in most cases are used as representing the catches of fish, are not very accurate and include much estimation. Furthermore, the space and time scales of reporting of landings do not necessarily coincide with the space scales of stock distributions or with ocean climate fluctuation scales.

The climatic effects on fish stocks are often assumed to be found in recruitment variations. Although climatic fluctuations might affect fish stocks by influencing recruitment, these effects are not immediate in catches as most fish biomasses are buffered by the presence of several year-classes in the exploitable part of the population. Recruitment variations are difficult to document and include many sources of errors, such as biases and errors in sampling of stocks and errors in age determination. Lapointe & Peterman (1991) pointed out that correlated estimates of abundance of recruits with ocean environmental variables and their anomalies have often been used by researchers interested in effects of climatic changes on fish stocks. However, they showed that this correlation is much dependent on the estimation of the natural mortality rate (M) used in virtual population analysis (VPA), and the estimation of M is always uncertain, especially when, as usually happens, there are time trends in the fishing mortality rate and in the environmental factors. Thus incorrect spurious correlations are created. Recruitment variations can also be affected by a multitude of factors other than environment, most of which such as predation, are found within the fish ecosystem. It is also very difficult to find the 'recruitment window' in time and space when and where successful recruitment originates (Laevastu *et al.* 1986).

Lundbeck (1962) has studied the effects of changes in fishing vessels and gear on catches and has converted the catches per unit effort to a given common basis (Fig. 1.1). He showed in this figure the changes of catches

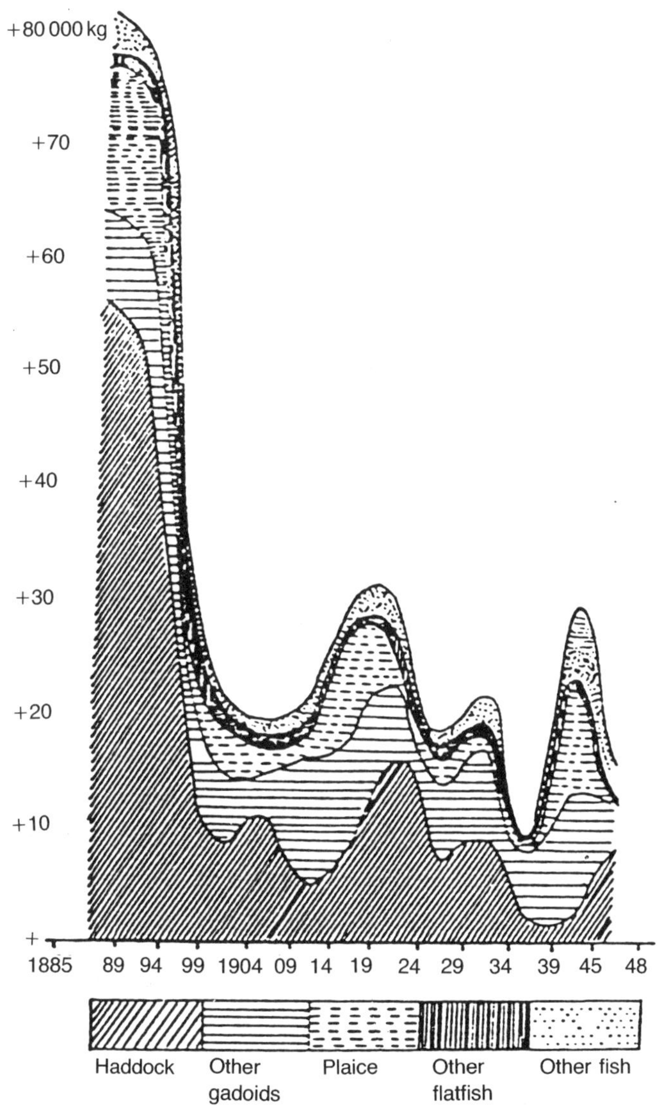

Fig. 1.1 Catch per 10-day trip (corrected effort) of a standardized German trawler in the southern North Sea (from Lundbeck 1962, Hempel 1978a).

and their shift between species. The high catch rate at the end of the last century was mainly caused by abundant haddock, which is known to have periodically very strong year-classes. Furthermore, Lundbeck was also able to show the areal shifts in distribution of species (see further Chapter 5).

Species dominance and replacement

Many apparent species replacements have occurred in different parts of the world (Daan 1980), where one dominant species declined and another ecologically similar species became dominant. Examples of these replacements are sardine–anchovy in California and herring–blue whiting in North Atlantic. It is difficult to determine whether food competition, oceanographic conditions, or fishing has played the primary role in causing these replacements. Replacements can be considered as normal fluctuations in the ecosystems, which are food limited and predation controlled. Thus the replacement phenomenon can occur also between demersal and pelagic fish, and such replacements are cyclic, e.g. cod–herring in the Baltic and partly in the North Sea. Considerable predation by herring on demersal eggs of cod has been observed (Daan 1976), and this might be one of the principal causes of apparent pelagic–demersal species replacement. Large and variable cannibalistic predation on herring (Johannessen 1980) or predation by jellyfish (Möller 1982) might also be among the causes of quasi-periodic fluctuations of this species. Cannibalistic predation alone can cause large cyclic variations in some dominant species such as walleye pollock in the Bering Sea (Laevastu & Favorite 1976).

Species replacement in some regions can be caused by changes of seasonal migrations without change of biomass. One such replacement has been described by Lluch-Belda *et al.* (1989) between the Monterey sardine and the crinuda sardine (a thread herring). In the Gulf of California local upwelling and colder water correlate with higher catches per unit effort for the sardine, whereas in warmer water the catch per unit effort of the crinuda sardine is higher. This phenomenon represents also a change of distribution.

Changes in distribution of species

Changes in the spatial distribution of a species are rather common, especially among pelagic species. The real as well as apparent change in distribution might be brought about by changing abundance of species in the ecosystem, by fluctuations of climate and by other factors such as selective fishing. An increasing population usually extends its area of distribution, whereas a decreasing population usually shrinks to the main centre of its distribution.

The effects of climatic fluctuations on the changing distribution of pilchard and herring along the south coast of Devon and Cornwall have been described by Southward *et al.* (1988). Pilchards are more abundant and extend further to the east when the climate is warmer. It is presumed that changes in climate can influence the relative competitive advantage of these and other species, and also have direct effects on reproduction and behaviour.

Local, minor climatic changes, e.g. in the position of oceanic fronts, can affect the distribution of feeding pelagic fish. Juvenile blue whiting have usually been found feeding on the warm water side of the polar frontal zone between Iceland and the Faeroes (Sveinbjörnsson *et al.* 1984). This frontal zone can shift somewhat seasonally and from year to year, thus causing changes in the distribution of feeding fish.

Changes in total finfish biomass

Until recently it has been assumed that the total finfish biomass in a given region is largely determined by the basic organic production (and other basic food, such as zooplankton and benthos) in that region. In addition, it was assumed that if the abundance of one species declined, other species would increase and fill the ecological niche, the total finfish biomass thus remaining fairly constant.

This concept of constancy of total finfish biomass has lately been challenged, first in upwelling regions, where it has been shown (e.g. by Shannon *et al.* 1984) that catches of short-lived pelagic species can be affected by the intensity of upwelling. Smith (1978) has pointed out that pelagic shoals in upwelling regions have to migrate in order to obtain enough food. Thus the change of biomass in a given region might be determined by immigration and emigration in search for abundant food. The competition between fish is at shoal level rather than species level (Smith 1978).

The great changes in fish ecosystem during the late 1980s in the Barents Sea have shown that the total finfish biomass can be dependent on the biomass of the main forage species, as about one-third of the food of commercial species consists of other fish (forage species). Drastic decline of the capelin (and earlier herring) biomasses also caused great decline of the cod and haddock biomasses in the Barents Sea, mainly because of lack of forage fish as food. The decline of the principal forage species can be caused by heavy fishing, increase of predator species and regional climatic fluctuations affecting recruitment and/or production of basic food (plankton). Thus, the decline of commercial species, such as cod, has often been directly caused by fishing and starvation with its consequences, but the principal cause, i.e. decline of forage species, might have been due to climatic fluctuations among other factors. Direct data are seldom available to demonstrate the principal causes and all the intervening processes.

The biomasses of commercial crustaceans can also fluctuate owing to seasonal and climatic fluctuations of environmental conditions. Dow (1964) has concluded that the temperature before and during the spawning of the Maine lobster, sea scallop and northern shrimp is correlated with, and might influence the abundance and availability of, these species and their recruitment in inshore waters.

Changes in landings caused by economic and technological factors

Various economic factors and technological developments affect the catches and landings of different species. However, it is difficult to quantify these economic effects, and to determine quantitatively their influence in the past records. The main economic factors affecting catches are:

(1) Changing market demands for given species. This demand also includes import–export conditions and distribution between different markets, such as processing and fresh marketing.
(2) Price change, including relative prices of species. An increase in the price of one species, may lead to more fishing effort being spent on catching this species.
(3) Changes and technological improvements in fishing fleets — number of vessels, vessel size and efficiency. These changes may also enhance the possibilities of exploiting more distant grounds. Changes in fuel prices affect the rentability of the fishing vessels and can cause changes in the mode of operation.
(4) Technological improvements in fishing gear, especially improvement of efficiency.
(5) Various management actions dictated by economic pressures from interest groups. There are now numerous such regulations, international as well as national, and more are coming.

All these factors listed above will affect the landings and will cause great difficulties in quantitative evaluation of the possible effects of climatic changes on fisheries.

1.3 Man-made and natural fluctuations of stocks

Superimposed on the fluctuations in abundance and availability of stocks caused by climatic fluctuations and economic-technological factors are the changes of stocks caused by man and man-made environmental changes. Thus, in order to evaluate the effects of climatic changes, we must quantitatively evaluate and separate the man-made effects whenever possible.

Man-made changes in environment and stocks

The greatest man-made change of fish stocks is caused by fishing. Fishing decreases the exploitable part of the stock, thus lowering the catch per unit effort in little shoaling species. Fishing alone seldom decreases the spawning stock to such a level that recruitment is materially affected. Recruitment overfishing is very rare indeed. In species which are predominantly highly cannibalistic, fishing can even promote recruitment by lowering predation pressure on pre-fishery juveniles by removing larger, more cannibalistic specimens. If fishing and other factors which affect the stock were to remain constant, a periodic fluctuation of the stock size would be created in highly cannibalistic stocks (Laevastu & Favorite 1976).

Despite the improving availability of data on catches and on the state of the stocks it is not becoming easier to separate the effects of fishing from effects of environmental (climatic) changes. Radovich (1982) pointed out that opposite views still exist on whether the collapse of the huge California sardine fishery was caused by overfishing or by climatic changes, despite a vast amount of past research on the subject. General effects of fishing on stocks are described in Chapter 5 Section 1.

Pollution of coastal waters resulting from man's activities, and associated auxotrophication (eutrophication), cause changes in the environment which in turn might affect the abundance of fish. Eutrophication in the Baltic Sea has caused considerable increase in production of pelagic species (herring and sprat) (Elmgren 1984). Elmgren (1989) estimated that the pelagic energy flows in the Baltic have increased by 30 to 70% in this century and the fish catches have increased more than tenfold.

The North Sea has been receiving from surrounding heavily populated countries wastes including heavy metals, PCBs and oils. According to Lee (1978) there is, however, no evidence that these pollutants have affected the well-being of the fish stocks. This is mainly due to better flushing of North Sea as compared to the semi-closed Baltic Sea.

Among other man-made changes around the North Sea and elsewhere are sand and gravel operations and the closing of tidal flats, which might take away fish nursery areas. The anadromous fish stocks (mainly salmon) are at present largely controlled by man, by changing the spawning grounds in rivers and/or by enhancing the stocks by setting out fry and smolts.

Changes within the fish ecosystem

Changes in the fish ecosystem, i.e. changes of abundance and/or distribution of species, can also occur with or without external influence. Some of these internal changes in the ecosystem are caused by competition for food and/or by predation, and might also be triggered by fishing. Andersen & Ursin

(1977) believed that the reduction of herring and mackerel in the North Sea in 1960s caused a gadoid outburst by diverting more food to gadoids. Cushing (1980) could not accept this thesis, as food distribution between species cannot be verified. He believed that climate deterioration might have been the cause, but could not verify that more food was made available for cod by the climatic change. Additional details of the changes within the marine fish ecosystem are given in Chapter 3, section 6.

1.4 Search for the causes of fluctuations

Many factors may affect the fluctuations of fish stocks, and the search for the causes of these fluctuations (and especially the quantification of the resulting effects) is not simple; it is impossible to be certain in most cases. Climatic causes have often been taken as an easy, non-verifiable cause, but this explanation often fails the test. For example, the changes of pelagic fish stocks in the North Sea have been assumed to be caused by changes in hydrographical conditions, especially in the flow of Atlantic water into the area, but this cannot be verified (Corten 1990).

A re-examination of past research into the effects of climatic changes on the fisheries and on marine resources is indeed called for. First, it is necessary to consider the basic data used in this research — landings and surface climatological and hydroclime data. Over 30 years ago Bell & Pruter (1958) suggested that a need for re-examination of the basis for some reported climatic temperature–fish productivity relationship exists. No adequate allowance had been made in the past for changes in the amount of fishing, economic conditions, and efficiency of the fleet, which are often considerable factors in fluctuations of landings.

Large scale alterations in the distribution of species may be caused by climate fluctuations, among other possible causes. However, we still do not know with certainty the causes of disappearance of the Bohuslän herring, although several hypotheses on this phenomenon exist. The decline of the Atlanto-Scandian herring and its change of life cycle migration routes are attributed to two plausible causes, climatic fluctuation north of Iceland accompanied by a decline of planktonic food, and heavy fishing on juvenile herring along the northern Norwegian coast. The drastic decline of West Greenland cod has been in the past ascribed to well documented, rather profound climatic (temperature) deterioration in this region. However, the latest investigations tend to support the theory that the decline of the West Greenland cod stock was caused by changes in the wind-driven Irminger current, which transports cod larvae and pelagic juveniles from the Icelandic spawning grounds to West Greenland.

The great changes in the fish ecosystem in the Barents Sea in the late 1980s have taught us to look into nearly all essential processes in the marine ecosystem, especially predation. A further lesson has been that the traditional population dynamics approaches used in fisheries research are inadequate and do not indicate the real causes of stock fluctuations.

This book attempts to review first the possible mechanism of the effects of climatic changes on fish stocks. Thereafter some pronounced changes in stocks are reviewed and the plausibility of the effects of climate fluctuations on these stocks is indicated wherever it is warranted.

1.5 Changes of elements of climate

In previous sections brief notes were given on the multitude of factors which affect the catches and landings of fish. These are among the basic data for research on climate–fish interactions; climatic data are also needed. It is difficult to find proper climatic data and to conclude that a climatic fluctuation has occurred over a defined region which might have affected the fish stocks through various processes. This section gives brief notes on the nature of marine climatic data pertinent to effects on fish stocks. Chapters 2 and 3 describe the weather and climatic processes more fully.

Changes of meteorological climatic elements

The meteorological elements of main interest with respect to the ocean and its resources are surface wind, its speed and direction, and factors and processes which affect heat exchange between the air and sea (wind, cloud cover, air and sea surface temperature, and water vapour pressure of the air). Air temperature, on which there are plenty of data, is an indicator of many changes in the atmosphere. Precipitation is of interest in some problems relating to run-off. A number of dependent climate elements, such as storms and storm tracks, are also of interest.

The smallest time scale of practical interest in studies of climatic effects is the season, and timing of the onset of conditions typical for the given season. In some instances in the study of the effect of environment on fish, attention must also be given to short time periods which fall between typical weather and climatic time scales, such as the critical periods during the larval life of given species.

The interest in meteorological elements focuses on their anomalies and effects on the ocean. In the past considerable attention has been given to various climatic semi-periodicities and to possible external factors such as sunspots. Popular subjects of study have been various long-term trends, such as the cooling trend in the atmosphere in the last 25 years in the North Atlantic described by Cushing & Dickson (1976), and the recently publicized opposite trend, global warming. Another way to present atmospheric climatic fluctuations is by the use of various indices (Dickson & Lamb 1972).

Climatic fluctuations are not necessarily global, but occur in all space scales. The changes in some climatic elements can be opposite in adjacent regions. On a large scale the changes in the southern and northern hemispheres often compensate for one another (Mörner 1984a). In dealing with the effects of climatic fluctuations on fisheries we are mostly interested in regional scales, or in space scales which correspond to the distribution of a given major commercial fish stock.

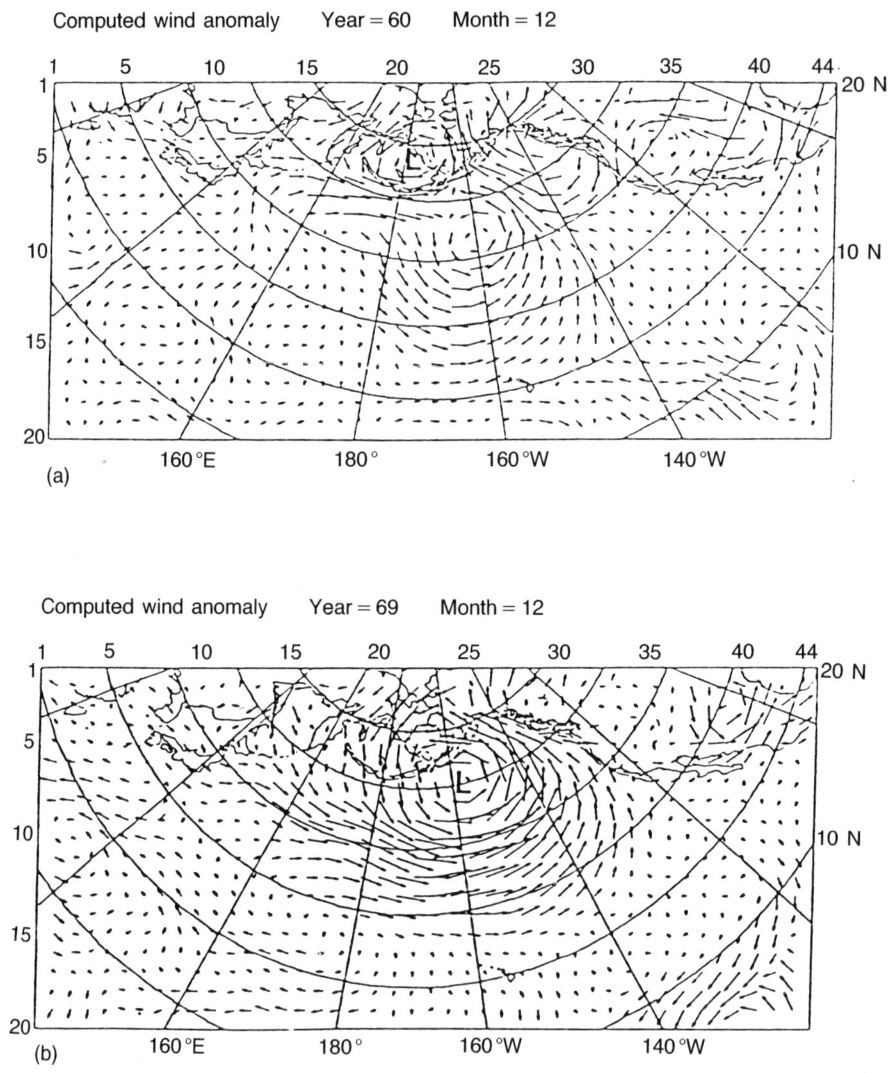

Fig. 1.2 Monthly mean surface wind anomalies for December 1960 (a) and 1969 (b) associated with low pressure centres (L).

Short term meteorological anomalies tend to have the dimensions of surface meteorological systems and are progressive in time. Monthly surface wind anomaly charts for the North Pacific (Fig. 1.2) illustrate the cyclical nature of these anomalies. Climatic anomaly data from a single meteorological station have been used in the past in correlation studies, and the question of how representative these data are must arise. Questions of significance of any statistically derived trend can also be raised in the light of the great season-to-season and year-to-year variations as exemplified by Stockholm's winter and summer temperatures from 1756 to 1983 (Fig. 1.3, Mörner 1984a).

Changes of ocean climate (hydroclime) elements

The changes in environmental variables in the oceans and their anomalies are caused by atmospheric driving forces and modified by processes such as mixing in the oceans. The ocean is an integrator of the effects of atmospheric forces; the changes in the oceans are usually slower and the anomalies are more persistent than in the atmosphere. The space scales of the anomalies

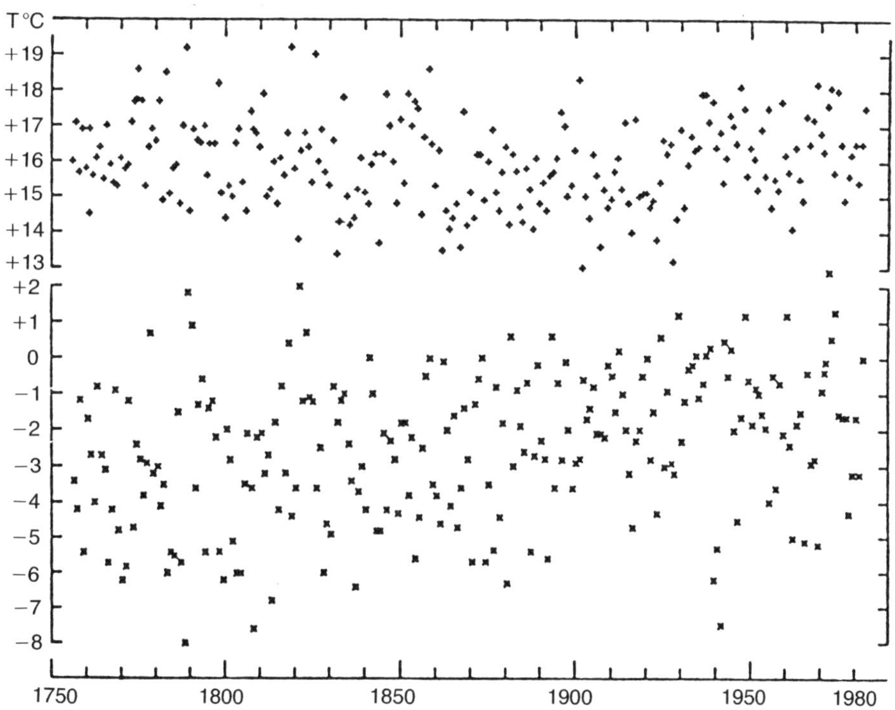

Fig. 1.3 Stockholm temperature records 1756–1983. Upper graph (+) gives mean summer (June–August) temperatures and lower graph (x) mean winter (December–March) temperatures (Mörner 1984a).

in the oceans are, however, usually smaller than the scales of large-scale atmospheric features and are also modified by ocean boundaries (e.g. the coastal upwelling). Long-term climatic changes and anomalies occur in surface layers, usually above 100 m depth; no long-term climatic changes are found below 200 m (Robinson 1960).

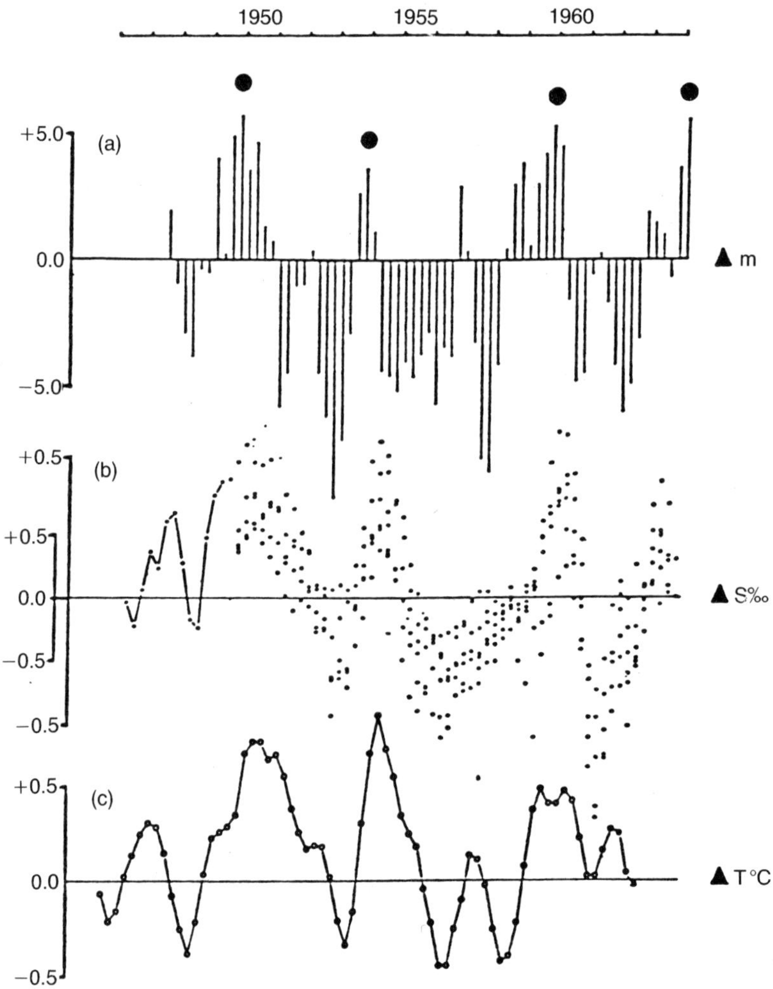

Fig. 1.4 Period 1946–63: Running 9-monthly means of: (a) The meridional index anomaly for the British Isles (southerly flow positive): (b) The surface salinity anomaly for the eastern North Sea (German Bight): (c) The 0–200 m temperature anomaly for the Kola meridian section (south-eastern Barents Sea). (b) is intended to show the trends of the salinity anomaly and not its actual magnitude. The two scales for the salinity anomaly arise because the diagram was obtained by superimposing two graphs, with different scales, derived from two sets of fields in the German Bight. Different scales were used because the amplitude of the salinity anomaly varied greatly from field to field. Both graphs had the same trends (Dickson & Lee 1969).

The extent and duration of ice are good indications of climatic fluctuations in higher latitudes. The maximum extent of ice in the Baltic is well correlated with mean air temperature. On the other hand, the maximum extent of ice is variable from year to year, with a slight long-term trend of decrease (see Chapter 3 Section 5). The extent of ice in the open ocean is in most cases related to wind, as shown by Malmberg (1969).

Wind causes currents and consequent advection of surface waters. Therefore, relationships between wind anomalies and changes of surface water properties are often found (Fig. 1.4). Large-scale surface wind anomalies and the anomalies of tracks of cyclonic systems (storm tracks) are in turn caused by displacement of large-scale surface weather systems such as the mean position and strength of the Icelandic Low.

Dickson & Lamb (1972) have described the atmospheric and oceanic anomalies in the North Atlantic, using various indices. The climatic changes in the North Pacific are less well studied. Favorite & Ingraham (1978) showed the surface pressure anomalies in the North Pacific to be related to the changing position of the Aleutian Low and ultimately the sunspot cycle. Ware (1986), using data from some coastal stations, was believed to have found 5- and 16-year periodicity in North Pacific ocean climate fluctuations off British Columbia and assumed these to be related to El Niño. However, questions can be raised about the validity of data from few single coastal stations as representing conditions in the ocean, as well as true periodicities in relatively short-term series observations.

Another peculiarity of oceanic anomalies is their slow progression from west to east, as reported by Rodewald (1966) and found also in the North Pacific (McLain & Favorite 1976). Similarly, the relatively slow eastward drift of 'The Great Salinity Anomaly of the 1970s' has been observed in the Atlantic.

The brief review of the complex problem of changes in fisheries and their possible relationships to climatic fluctuations indicates the need to look in greater detail at marine climatic changes, the effects of weather on the ocean, and the effect of changing ocean environment on fish. This attempt is made in Chapters 2 and 3.

Chapter 2
Marine Climate and Weather and Their Effects on Oceanic Conditions

This chapter summarizes the essentials of marine meteorology, marine climates, and the sea–air interactions, with emphasis on how weather affects the ocean and on those subjects which it is necessary to consider when evaluating the effects of weather and climate on fish behaviour and fishery resources in general. Furthermore, mainly medium and high latitude weather systems are considered. Detailed descriptions of maritime meteorology are found in a few books such as Roll (1965) and Palmen & Newton (1969). Weather forecasting at sea is described in WMO (1966).

Fishermen are interested in the present, near past and near future weather for tactical fishing operations and for searching for fishable aggregations of fish. Some past experiences and/or plausible hypotheses should, however, indicate that a relationship between weather and fish aggregation exists. The fisheries officials whose task is resource assessment and allocation are more interested in fluctuations of some climatic elements, hoping that some climatic influence on the stocks, especially recruitment, can be discovered and used as a predictor. However, a knowledge of how the climatic influence is transmitted to the stock is also useful. The mechanisms of the influence of weather and climate on fish and fish stocks are reviewed in Chapters 3 and 4.

2.1 Marine weather and climate and the cyclonic systems

The essential surface weather elements of concern to fisheries are surface winds (including storms), fog, ice and, only marginally, temperature and clouds. The most important are surface winds, which, though observed, are mostly computed from surface pressure distribution.

The atmosphere is in constant circulation around the Earth from west to east, most clearly at higher levels (e.g. 500 mb or about 5.5 km high). On the surface this circulation is more complex and is usually divided into three larger zones, the trade wind zone, the doldrums at the equator, and the horse latitudes at about 30 °N, prevailing westerlies which are bounded by the polar front south of 60 °N and the polar region. On the higher levels in the atmosphere wave-like motions develop in middle latitudes, and on the surface vortices occur which, according to their direction of circulation, are classified as cyclonic (circulation around low pressure) and anticyclonic (circulation around high pressure). The direction of the wind is not parallel to surface isobars, but under an angle towards lower pressure, the angle of deflection, which varies with latitude.

The main surface features at medium and high latitude are cyclonic systems. These cyclonic systems at the surface start as frontal waves, develop into mature systems and decay into occluded systems. The cyclones are associated with somewhat regular and predictable wind, air mass structure, cloud and precipitation cycles. The movement of the pressure systems on the surface is largely guided by the 500 mb patterns. The cyclones move from west to east along some patterns (Fig. 2.1), which vary from season to season and somewhat from year to year (Figs 2.2, 2.3). The strongest winds

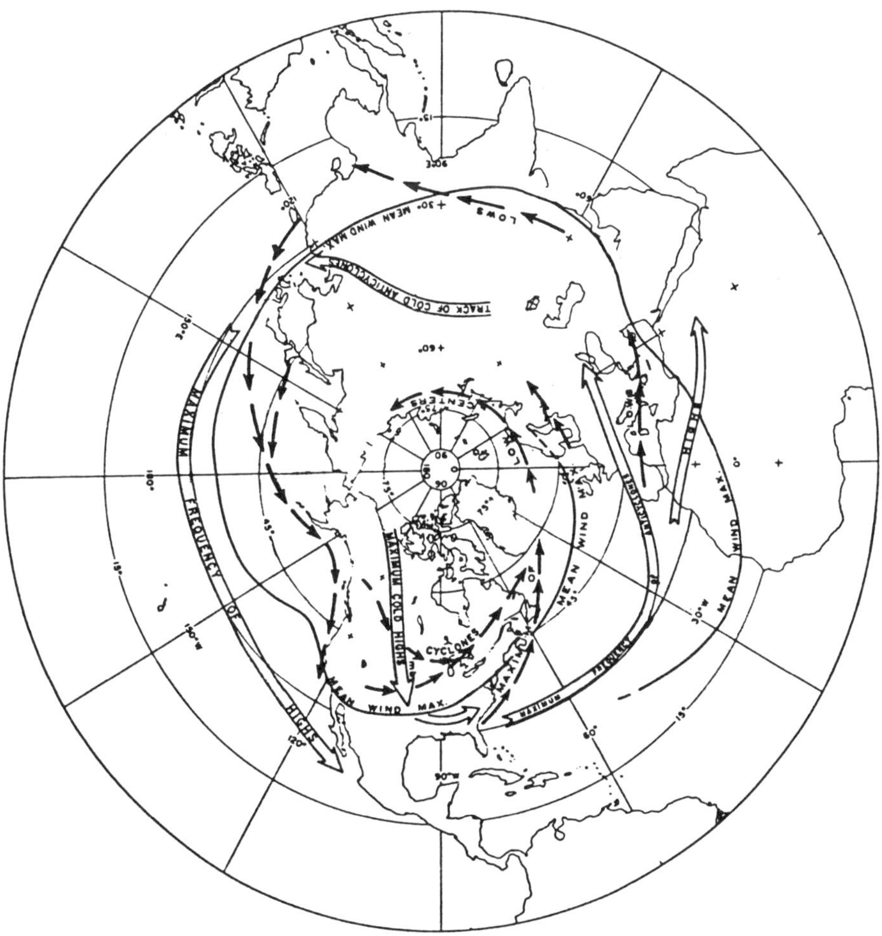

Fig. 2.1 Axis of mean maximum wind (heavy line) and simplified axes of maximum frequency of occurrence of cyclones (short arrows) and anticyclones (double-shafted arrows) for the winter season (Palmen & Newton 1969).

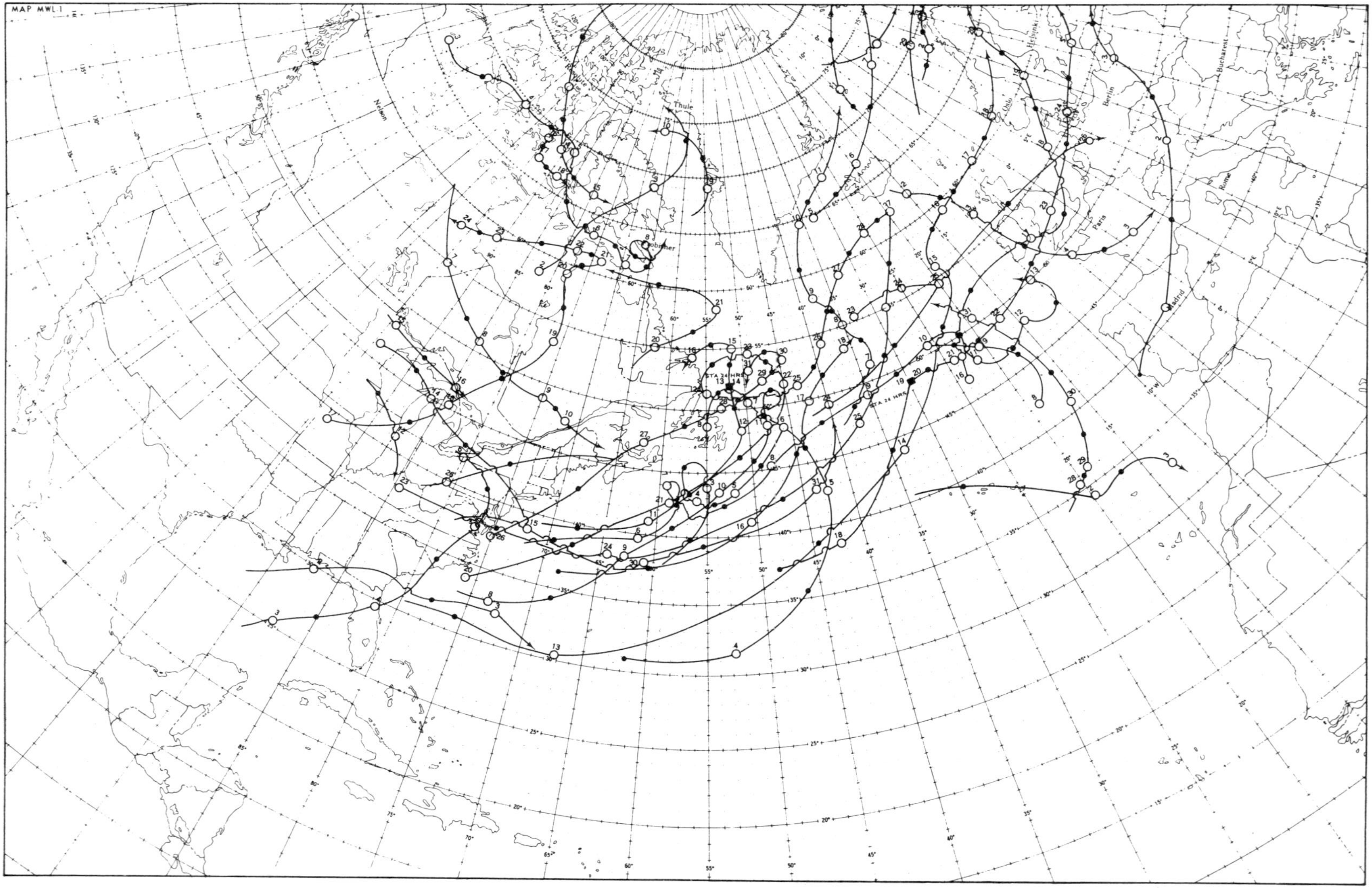

Fig. 2.2 Storm tracks of centres of cyclones at sea level in the North Atlantic, March 1965.

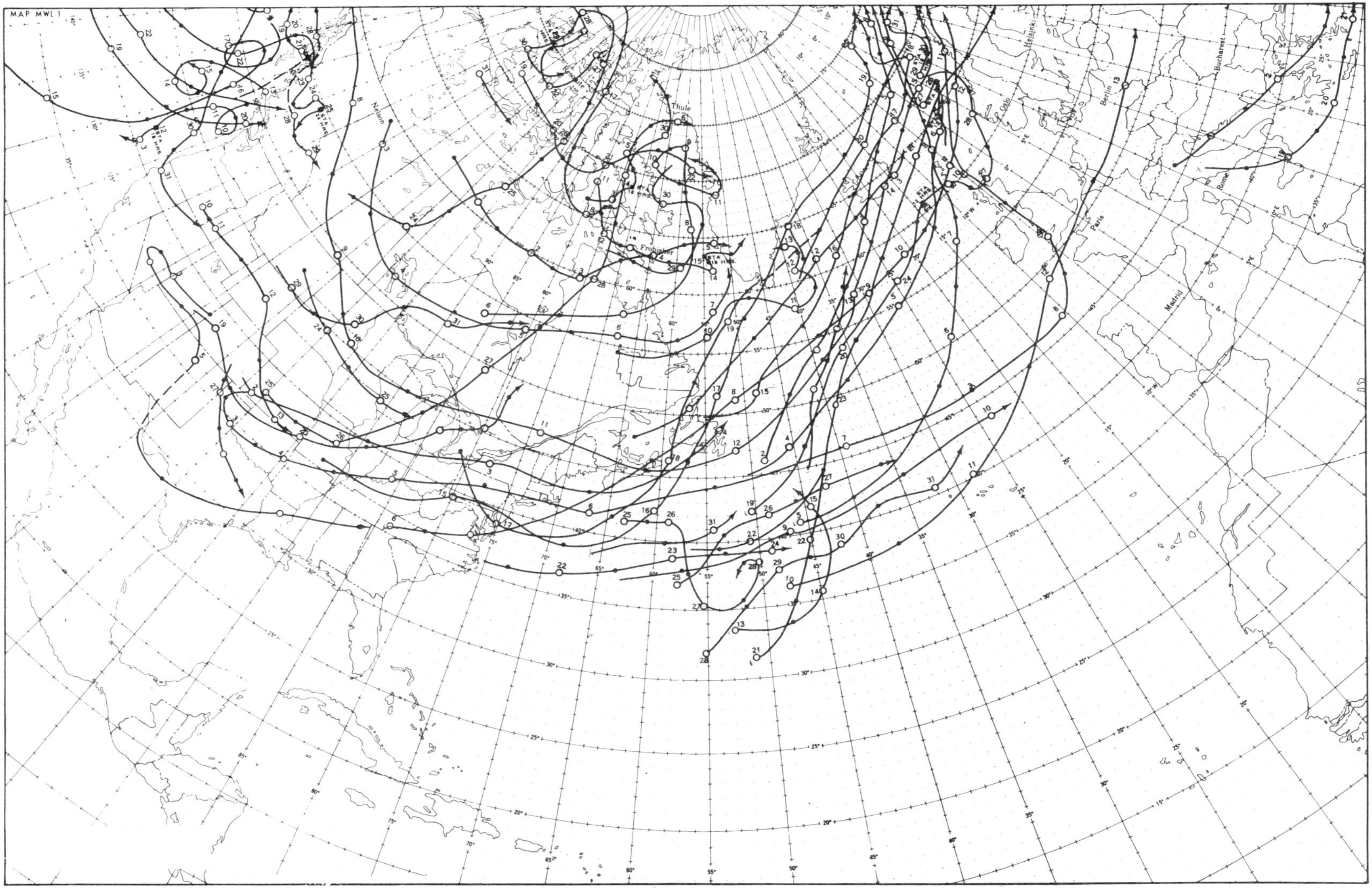

Fig. 2.3 Storm tracks of centres of cyclones at sea level in the North Atlantic, March 1967.

are associated with the cyclone tracks. Therefore, the year-to-year anomalies of the storm (cyclone) tracks are one of the main causes of anomalies in the oceans. These anomalous storm tracks are also depicted with surface pressure anomalies.

The average state of the weather is the climate, which can be characterized by the periods of summation of the mean elements (usually monthly or seasonally). When daily surface pressure patterns are summarized to form a monthly mean, some organized smooth pressure patterns result (Figs 2.4, 2.5), where some special features can be recognized, such as the mean Icelandic Low and the Subtropical High in the Atlantic (Fig. 2.4). There are corresponding features in the Pacific Ocean (Aleutian Low and Pacific Subtropical High). These figures also show mean wind patterns which are also variable with seasons and latitudes.

Climatic fluctuations are well described by the deviation of the mean surface pressure features (e.g. displacement of the Aleutian Low in a given month) and/or by computing the surface wind anomalies (Fig. 1.2).

The main changing features in lower latitudes are the variations of the strength of trade winds, changes in monsoons, and the occurrence of tropical storms (hurricanes, typhoons, and cyclones). The tropical storms, which are associated with storm surges, are of concern to navigation, offshore fisheries, and coastal areas of their landfall.

A special feature of interest in the atmospheric circulation is the formation of persistent highs in 500 mb (Fig. 2.6) which form predominantly during winter and spring and can persist up to several weeks. These blocks form most frequently over the Gulf of Alaska and the southern Norwegian Sea (Fig. 2.7), which are areas where cold air flows over warm water. The surface systems approaching from the west are forced mostly to go north of the block; only a few pass south of it (Berry *et al.* 1954).

The conversion of surface pressure fields to wind fields is a necessity, but not always a simple task, as both speed and direction of wind must be resolved. This complexity is illustrated in Figs 2.8 and 2.9 (Verploegh 1967).

The common pressure and surface wind anomalies as caused by north–south displacement of Icelandic and Aleutian Lows are illustrated in Figs 2.10 and 2.11. Fig. 2.10 shows a pronounced pressure anomaly (negative) in the North Pacific which is caused by southward displacement of the Aleutian Low. The Icelandic Low is displaced towards the north and west (positive pressure anomaly). The corresponding wind speed anomalies are given in Fig. 2.11. In the North Pacific the negative wind speed anomaly occurs in the northern part and the positive speed anomaly (increased wind speeds) in the southern part of the pressure anomaly.

Two characteristics of wind anomalies could be emphasized, which should be borne in mind during evaluation of the effects of climatic fluctuations on the ocean and on its resources. Anomalies are mostly regional and can

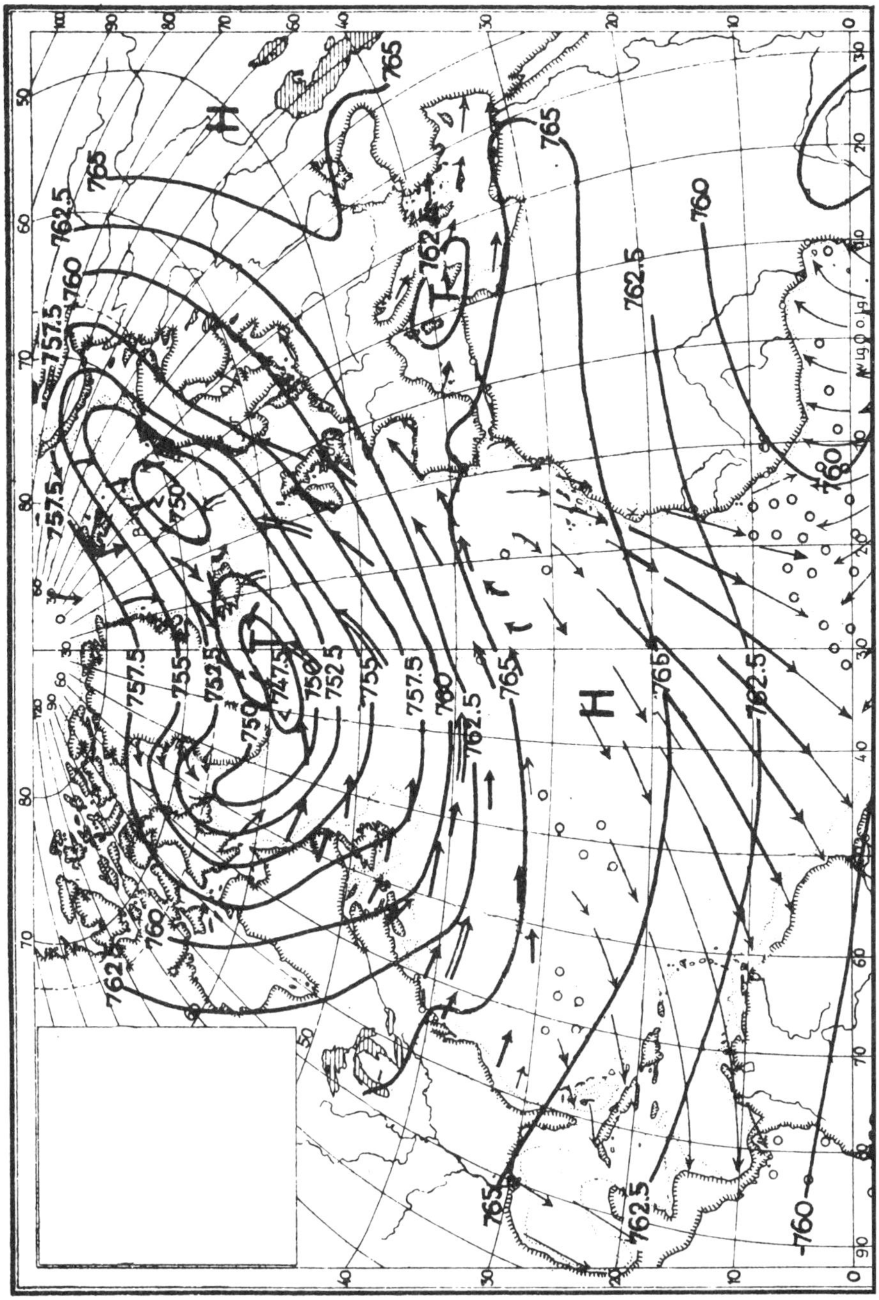

Fig. 2.4 Monthly mean surface pressure (in mmHg) and winds over the North Atlantic Ocean in February (Schott 1944).

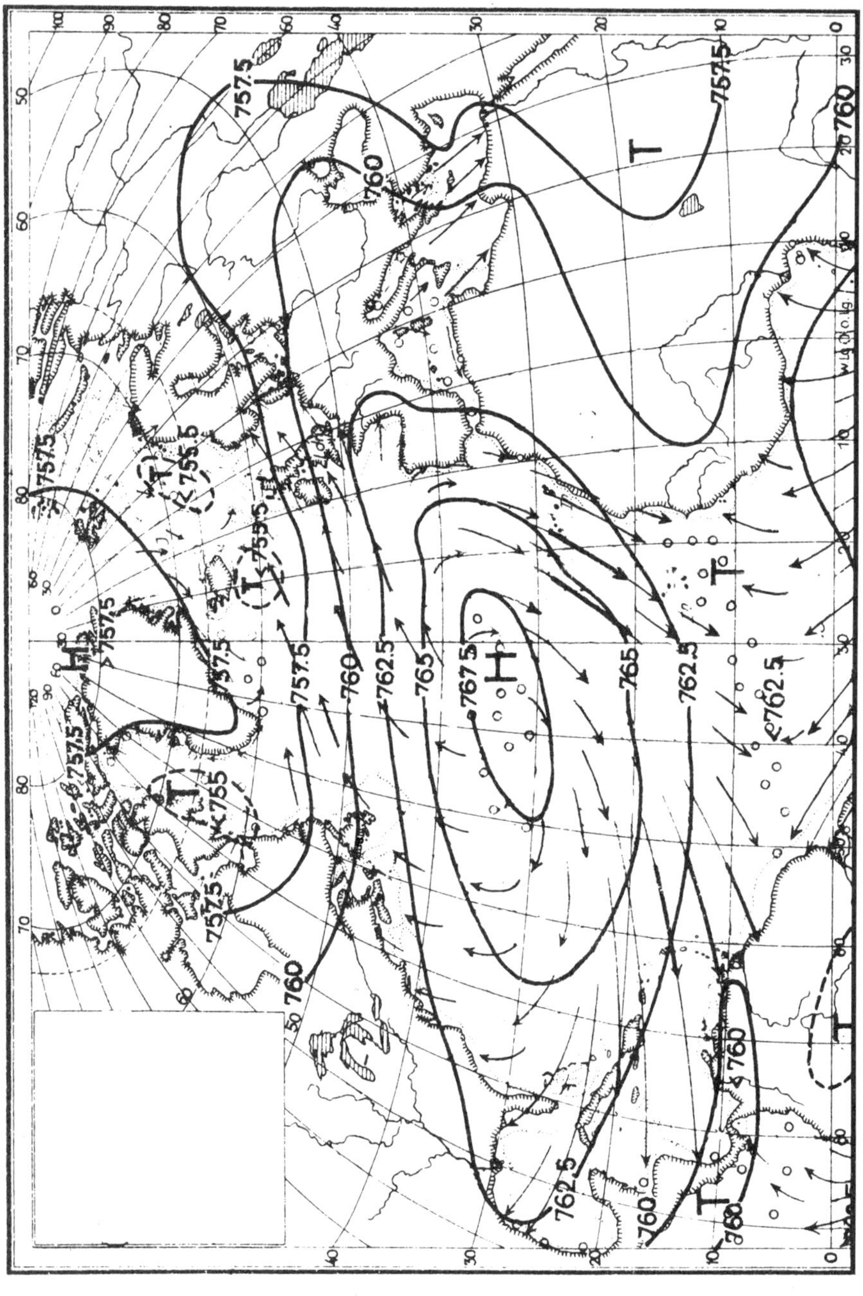

Fig. 2.5 Monthly mean surface pressure (in mmHg) and winds over North Atlantic Ocean in August (Schott 1944).

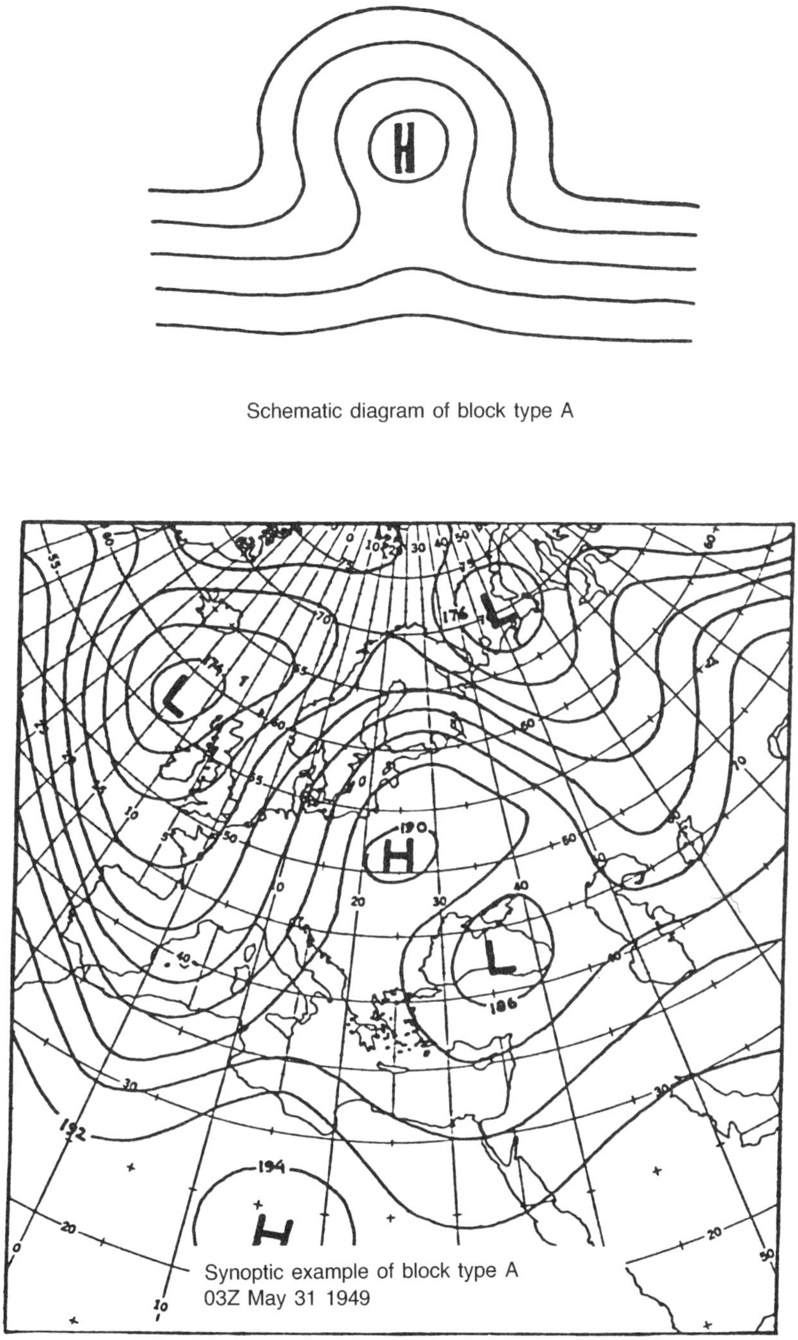

Fig. 2.6 Schematic diagram of block type A and synoptic example of block type A at 500 mb, May 31, 1949 (George & Wolff 1955).

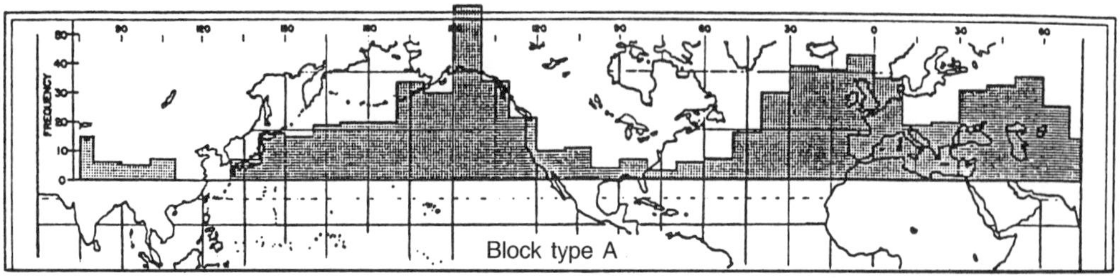

Fig. 2.7 Longitudinal distribution of daily block occurrence (George & Wolff 1955).

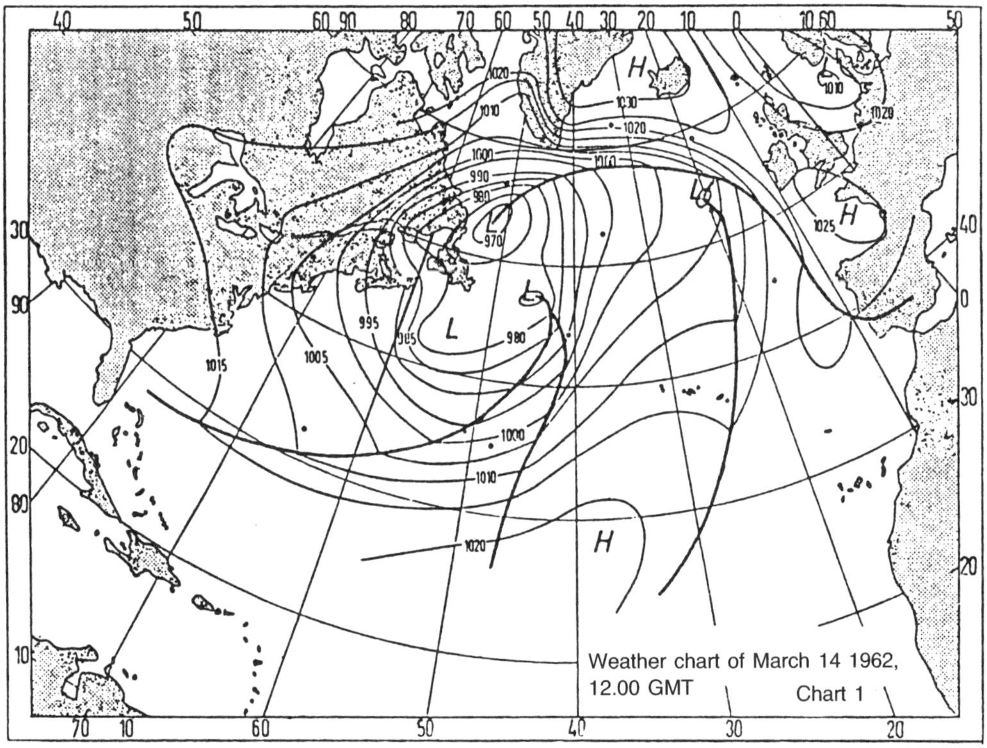

Fig. 2.8 Weather chart of 14 March 1962, 12Z, surface pressure (mb) (Verploegh 1967).

sometimes be quite local (e.g. limited to coastal areas). Anomalies (deviations) in one region are usually accompanied by changes in the opposite direction in neighbouring regions.

Some of the long-lasting climatic wind anomalies have been described briefly in numerous articles on climate fluctuations (e.g. Lamb 1984a).

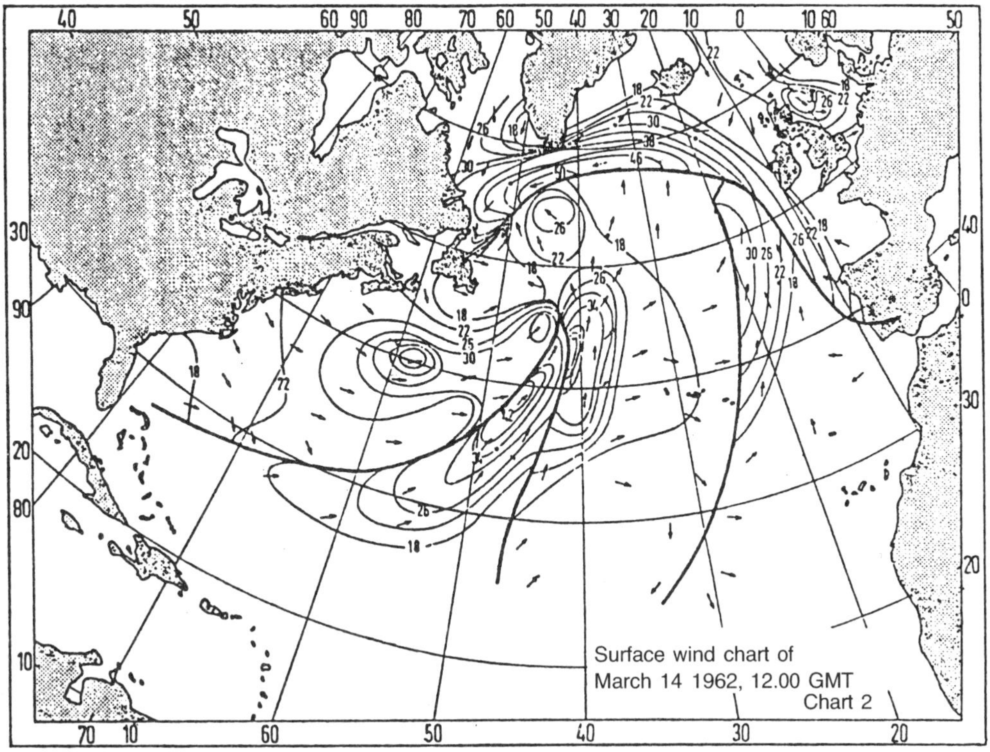

Fig. 2.9 Surface wind chart of 14 March 1962, 12Z (isotachs in knots) (Verploegh 1967).

Causes of the anomalies and connections between them have also been sought and described. Dickson & Namias (1976) have described the pressure and wind anomalies in the Iceland–Greenland area in the mid 1950s and the early 1970s as being caused by anomalies along the northeast coast of North America. These teleconnections are thought to rise through the influence of the ocean on the atmospheric circulation effected through the energy exchange between the sea and the air (Tokioka 1983).

2.2 Weather forecasting, forecast availability, and accuracy

The fishing industry is dependent on good weather forecasting, for safety and comfort as well as for foreshadowing availability of fish to gear. In turn, offshore fishing vessels can, and should, provide synoptic weather observations for the initial analysis of weather upon which the forecasts are based.

Weather forecasts are now prepared numerically in national and a few intergovernmental weather centres, such as the Medium Range Weather Prediction Facility in Reading, UK, and disseminated by radio and facsimile

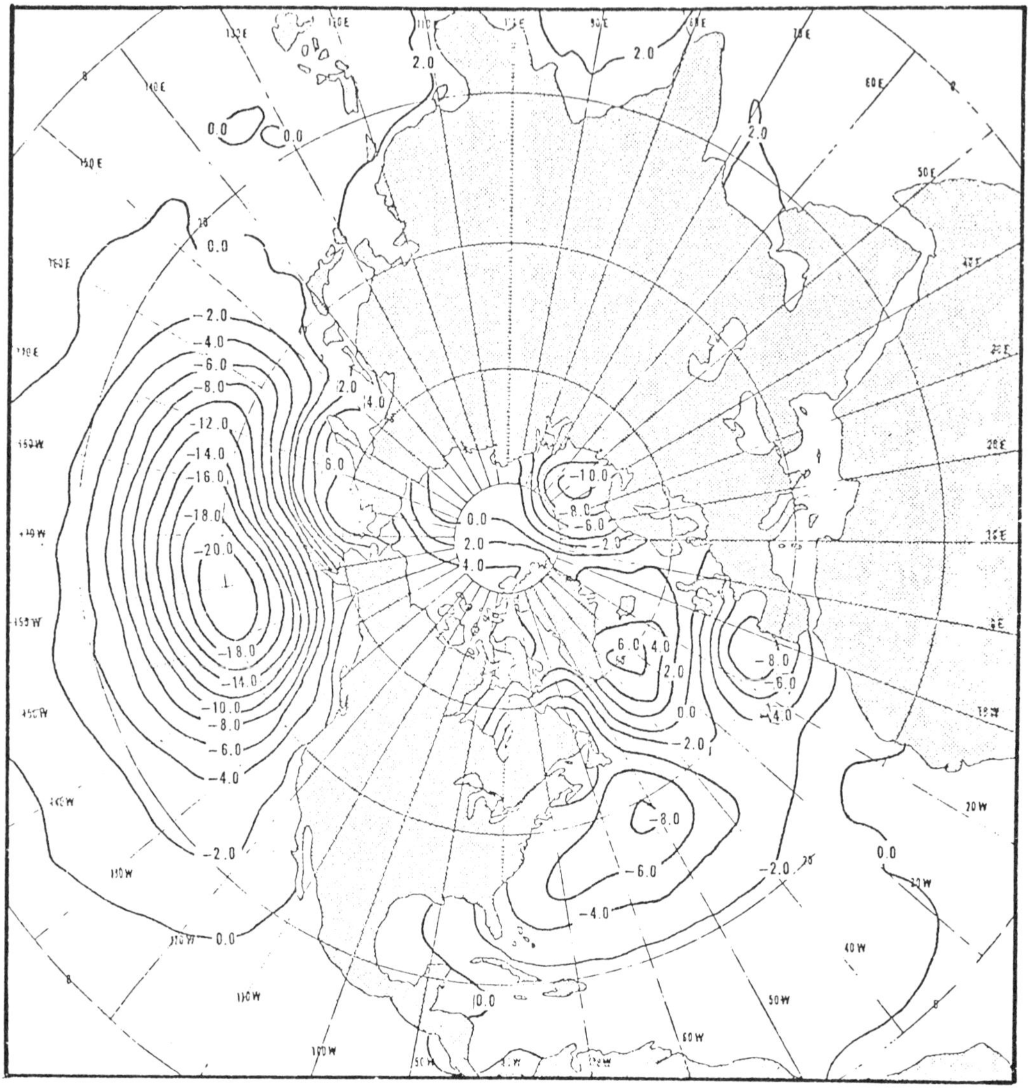

Fig. 2.10 Monthly mean surface pressure anomaly (mb) in February 1968.

to ships and coastal stations. In the following subsection only a few essentials of weather forecasting are presented, extracted mainly from WMO (1966).

Weather information is broadcast to shipping in weather bulletins which contain the following parts:

(1) Storm warnings in plain language.
(2) A plain language summary of the synoptic weather situation.
(3) Plain language forecasts for the areas of concern.

Fig. 2.11 Monthly mean wind speed anomaly in February 1968 (m s^{-1}).

(4) Synoptic weather map analysis in the abbreviated International Analysis Code for marine use (IAC FLEET), WMO Code FM 46.C.
(5) Selected ship reports in WMO Code FM 21.C.
(6) Selected land station reports in WMO Code FM 11.C.

Parts 1–3 in plain language are very familiar to most fishing skippers and other mariners. Parts 4–6 are the ones which will provide the mariner with more detailed information. However, they require plotting of the weather map at sea, for which fishing vessel mates scarcely have time.

Parts 5 and 6 of the weather bulletin are usually transmitted ahead of Part 4 in the schedule because the reports themselves are available before the analysis which is made from them. Many countries are assigned weather forecast areas by WMO and are broadcasting fax as well as radio forecasts in their languages. Most of the forecast areas are relatively large and the forecast validity extends up to 24 h. Weather conditions frequently change considerably in space and time and the value given in the forecast represents average conditions for the area and forecast period. Information on transmission schedules is found in WMO Publication 9, Vol. D. Most offshore fishing vessels carry fax and can receive analysis and forecast charts.

The interpretation of analysis and prediction charts requires local (regional) knowledge and experience of the peculiarities of weather in different regions. Some of this knowledge is available in Pilot Books (Sailing Directions).

Fog, which is one of the great dangers besides storms to navigation and fishing, occurs frequently in some coastal areas with upwelling as well as at the boundaries of cold and warm currents. There are some simple forecasting rules for fog which require the forecast of wind speed and direction and sea surface temperature gradient.

Weather systems move and undergo development in size, shape, and intensity. The greatest changes of surface conditions occur at fronts. The speed of movement of the system depends on the synoptic situation and season. A frontal cyclone type moves at 18 to 25 knots, whereas an occluded (filling) system moves at only about 10 to 15 knots. The development and movement of the weather systems depend on the large-scale weather patterns as well as on the types of gales and storms.

Some general weather forecasting rules exist which are, however, simplifications and their validity depends on many conditions. Some of these rules, especially those which might be useful for interpretation of near future fishing conditions as well as fish availability, are given below:

(1) The rule of persistency — the rate of movement and change of intensity continues as shown in the recent past.
(2) When a depression has a large open warm sector, a deepening of the depression is to be expected, the speed of which usually increases with the narrowing of the warm sector; it decreases, however, when the occlusion process is going on.
(3) Large depressions, when totally occluded, move very slowly and sometimes irregularly.
(4) Small depressions caught up in the circulation of a larger system have a movement following the main circulation.
(5) The more a secondary low deepens, the more it approaches the centre of the primary depression. Eventually it will absorb the old primary and become the primary itself.

(6) The speed of a front is largely determined by the force of the component of the wind at right angles to the front. A front moves faster the more the pressure falls before it in the case of the warm front or the more the pressure rises behind it in the case of the cold front.
(7) Frontal depressions tend to occur in families, each depression following approximately the path of its predecessor but displaced somewhat towards a lower latitude.
(8) When a deepening frontal warm sector depression moving along the east coast of a continent is approached by the cold front of another, continental, depression, a severe cyclone is likely to develop within 24 h.

More exact methods for predicting movement and changes of intensity of cyclones have been worked out in the past (e.g. George & Wolff 1955), which would be useful in a forecasting office. Nowadays all weather analysis and forecasting is done on large mainframe computers. However, the interpretation of the analysis/forecasts for the limited local area must still be done on board the vessels. Large scale and significant weather systems can now be computed with reasonable accuracy up to 5 days ahead, and with improvement of resolution and realistic parameterization in the future can produce accurate forecasts up to 10 days (Bengtsson 1981). Any weather forecast beyond this time limit has no value, although monthly and seasonal forecasts have been, and are made.

Surface weather analysis and prediction are mostly done in terms of surface pressure distribution. This distribution must be interpreted to wind speed and direction for practical use at sea. Geostrophic wind speed and its cross-isobar angle are obtained from surface pressure distribution. Furthermore, this geostrophic wind speed must be converted to true surface wind speed and direction, for which instructions and scales are provided on weather plotting charts. For more accurate work several other factors besides isobar spacing must be used, such as stability conditions (Hasse 1974). For rapid manual conversion on board ship simple methods are adequate, using tables or diagrams and wind nomograms (e.g. Rudloff's nomogram).

In estimating the wind force from the weather map plotted aboard ship the following should be taken into consideration:

(1) Anticyclonic or straight isobars produce higher wind forces than cyclonic isobars of the same spacing.
(2) Low latitudes produce higher wind forces than high latitudes for the same spacing of isobars.
(3) Cold, unstable air masses over warm water produce higher wind forces than warm, stable air masses over cold water, for a given isobaric spacing.

To estimate the wind direction on the open sea from a weather map, it is sufficient to assume that the wind crosses the isobars at an angle of 10 to 20° toward lower pressure. However, with light to moderate winds and at lower latitudes greater deviations, up to about 45°, are found.

Near the coast, especially near elevated coastlines, it is more difficult to estimate the wind direction from a weather map. In some coastal waters there is a distinct tendency for the wind to blow nearly parallel to the coast, in either direction. Quite near the coast, downslope winds of the fohn or bora type may blow nearly at right angles to the isobars and to the general direction of the coastline.

Sea fog is caused primarily by warm air flowing over relatively cold water. Areas or broad tongues of cold water surrounded by warmer sea or land areas are particularly favourable for fog formation. Such cold water areas may result from advection through polar currents (e.g. the Labrador Current), from the upwelling of colder water near certain coasts, from vertical mixing by tidal currents (e.g. in the English Channel), or from differentially greater seasonal heating of surrounding land areas (e.g. the Baltic Sea). The risk of fog varies with the season, the direction and force of the wind and other factors.

Heavy ice accretion on the superstructure of ships occurs when the air temperature is well below freezing point and winds of gale force or higher are blowing. Sea surface temperatures near 0°C favour ice accretion from spray. Smaller ships, such as trawlers, are in danger of capsizing from severe icing, but icing can be hazardous also for larger ships.

Accurate numerical weather forecasts beyond 7 to 10 days are at present impossible, mainly because of turbulent bursts in all time and space scales. Some foreshadowing of weather has been attempted, using a great variety of approaches. According to Flohn (1965):

> 'Practically all statistical techniques are designed to forecast not an individual (perhaps extreme) case, but a most probable sequence of events; they smooth the real weather trend and suppress each unusual development. Therefore the practical value of statistical long range forecast techniques remains limited. Such procedures are frequently misleading, as long as they are based on more or less randomly selected parameters instead of rational physical relations'.

Large and long-lasting weather anomalies have been observed to be correlated with time and area anomalies of the general atmospheric circulation. These anomalies occur in some regions and can be opposite in sign in other regions. Study of these anomalies has led to attempts at forecasting with teleconnections. Empirical studies on large scale teleconnections have revealed some physical inter-relationships, some of them acting nearly simultaneously, some with a time lag apparently suitable for long range forecasting. However, attempts to use these teleconnections for forecasting have not given the desired results.

Sawyer (1965) pointed out that despite the manifest importance of weather anomalies which last for weeks or months and the active attempts which have been made to solve the practical problem of long-range forecasting, little is conclusively known about the physical factors which cause long-term weather anomalies. It can be stated that despite considerable effort spent on long-range forecasting in the past, no useful and reliable method is at hand.

2.3 Sea and swell

Waves in the sea, generated by local and distant wind fields (wind waves and swell, respectively), are the most significant phenomena at sea which affect safety, comfort, fishing operations, and fish behaviour and availability. There are three different effects of waves on the sea below the surface which might be of concern to fisheries:

(1) Vertical mixing by wave action and turbulence caused by breaking waves. This wave mixing can deepen the surface mixed layer depth and 'sharpen' the thermocline gradient. Furthermore, it can affect fish directly by making them 'seasick' and inducing them to move deeper, where the orbital movement of water, caused by waves, is absent.
(2) Waves cause current (mass transport by waves) in addition to surface wind drag.
(3) Breaking waves cause 'wave noise' which might affect fish behaviour.

At any time there can be present waves of different height, period (length) and direction, caused by either local winds or winds from distant wind fields (swell). Usually one can recognize predominant waves and predominant swell, unless the sea can be termed as confused sea. The waves caused by the local winds and storms are usually relatively steep and short. The waves which travel out from the generation area are long and round.

A voluminous literature exists on wave theory, observation, analysis and prediction. Synoptic numerical wave analysis and prediction are now available for most sea areas. The following description of wave prediction presents only a few essential and simple facts, which might be useful for estimations of wave parameters in the absence of more accurate data.

Wave parameters depend not only on wind speed, but also on its duration, direction and change with time. The observation of wave parameters at sea is greatly dependent on experiences. The usually observed wave height is termed significant wave height. It can be expected that every tenth wave is about 1.5 times higher than the significant height.

The wave forecasting methods can be classified into the singular empirical methods and methods based on wave spectrum and statistics. The methods based on spectrum and statistics are time-consuming in use but are more flexible and suitable for construction of wave fields over the oceans.

Singular methods for wave analysis and prediction are based on empirical observations and on one or several of the relationships between wave height and/or length and wind direction, fetch length, sea-air temperature difference or season, depth of the water and other factors, but give only one statistical parameter of the waves, such as average and/or significant height. The accuracy of wave forecasts depends very greatly upon the correct estimation (forecast) of surface wind speed and partly on wind duration.

Simple singular empirical methods for wave height estimation are rapid in use and are useful for estimating the height of the fully grown sea in a few given locations on a weather map, and serve at times for auxiliary computations in other forecasts such as determination of depth of forced mixing in the sea.

Four different formulae for estimation of fully grown seas are compared in Fig. 2.12. Only one of them (Laevastu 1960) takes into account the sea–air temperature difference and can be used for estimation of non-fully developed seas, as it contains factors of wind duration and fetch length. Laevastu (1960) also gives an approximate empirical formula for estimating average wave length if the significant wave height is known. Length in metres equals 50 times the square root of the height in metres.

Lumb (1963) developed a simple method for forecasting maximum wave height in the North Atlantic based on the analyses of wind and wave data from weather ships *J* (52° 30′N, 20° 00′W) and *I* (59° 00′N, 19° 00′W). The three straight line relationships between maximum wave height and wind are given in Fig. 2.13.

It should be noted that line I on Fig. 2.13 is valid from February to April when the cold air prevails over the warmer water, and line II for the same period but with the air warmer than the sea surface. Line III is valid from May to August. Those straight lines are not extended below 15 knots and in the case of line II not below 20 knots. According to Lumb the reason is that with light winds swell often dominates over the sea and the simple relationship between maximum wave height and the average wind represented by the straight line then breaks down.

Rough estimates of wave conditions can be obtained by the use of an empirical table (Table 2.1), where the average wave height is related to wind force or speed. This table is intended only as a guide to show roughly what may be expected in the open sea remote from land. In enclosed waters, or when near land but with an off-shore wind, wave heights will be smaller and the waves shorter and steeper.

The singular approaches in wave forecasting are based on fully developed seas. Only about one-third of the seas are fully developed in a given wind

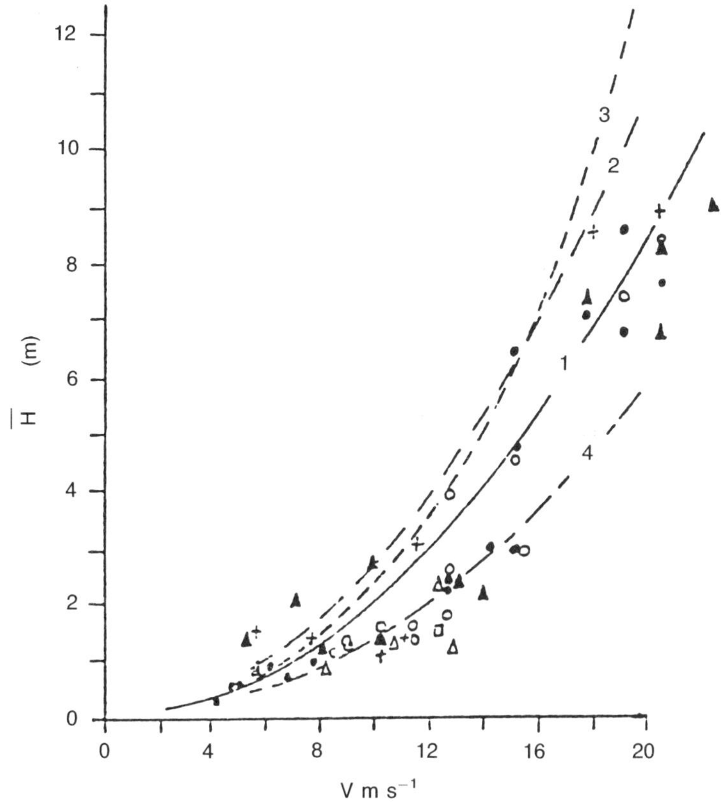

Fig. 2.12 Significant heights of fully developed sea according to Darbyshire (4), Neuman (3), Sverdrup-Munk (2) and T. Laevastu (1).

speed. A very approximate role for estimation of the change of wave height by decreasing wind can be drawn from the data presented by Walden (1959). Change of wave height in metres equals one-tenth of the change of wind speed in knots. The decay seems to be relatively rapid after the decrease of wind speed because the higher waves generated by the earlier higher wind move out of the area of generation.

It is well known to seamen that the currents appreciably affect the wave characteristics. If the current runs against the sea, the waves are steeper, the length appears shorter and the seas are very choppy. If the current runs with the sea, the seas are smoother and the waves rounder and longer. These effects can easily be observed where strong tidal currents occur in channels between islands.

There is a mass transport (wave current) connected with waves. Therefore, a long swell affects also the shorter superimposed waves in the same way as currents do. A swell which runs nearly in the opposite direction to the sea

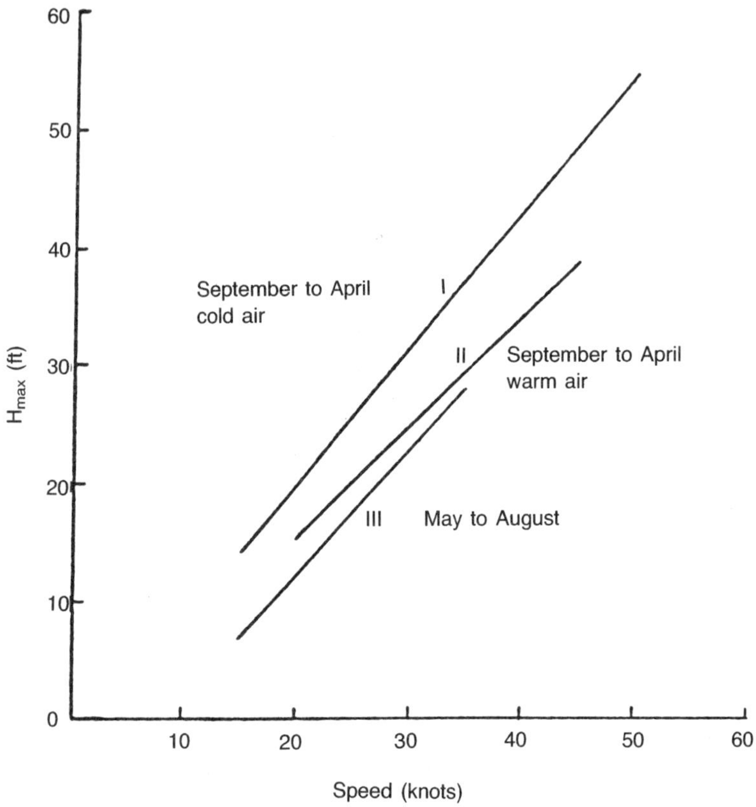

Fig. 2.13 Lumb's graph for estimation of maximum wave height from the mean wind speed over the preceeding 12 h (Lumb 1963).

Table 2.1 Approximate wave heights related to Beaufort wind forces 0–12

Wind		Probable wave height (m)	
Bft force	Speed (knots)	Average	Maximum
0	{1	–	–
1	1–3	0.1	0.1
2	4–6	0.2	0.3
3	7–10	0.6	1
4	11–15	1	1.5
5	16–21	2	2.5
6	22–27	3	4
7	28–33	4	5.5
8	34–40	5.5	7.5
9	41–47	7	10
10	48–55	9	12.5
11	56–63	11.5	16
12	64–71	14	–

causes steeper, shorter waves, and a swell running in the same direction as the sea causes the sea (wind waves) to be less steep, round, and longer.

Storm surges are caused by strong storms. Their effects are combined from wave effects (large waves, wave build-up), currents, and atmospheric pressure effects. They can be considered as events rather than climatological factors. Damage caused by them to local mariculture, coastal fisheries, and fishing boats in harbours can be considerable.

2.4 Surface currents and their variability

Currents and their changes are significant features in fishing operations, changes in fish availability and abundance, and in any possible climatic change in the oceans affecting fish.

Surface currents are typically variable in direction and speed, even in cases where the tidal components are subtracted and the local wind is steady. Stommel (1954), for example, investigated the surface currents off Bermuda using drifting wireless-telemetering buoys, and found that there was considerable irregularity in current speed and direction, even during days when the wind was fairly steady. He called this irregular motion, after Ekman, a kind of 'macro-turbulence'. Knauss (1960), using neutrally buoyant floats, also found erratic motion in deeper water below the thermocline (300 to 2140 m). The reasons for this variability of currents have been sought also in several other forces besides wind and tides which, independent of each other, affect the surface currents simultaneously. In addition, the conditions in neighbouring areas (piling up, differences in gradients of driving forces, etc.) affect surface currents in a given area and cause various eddies, both macro- and meso-scale as well as convergences and divergences.

Various approaches have been used in the past for prognostication of surface currents, but actual synoptic analyses and forecasts of surface currents have been very limited. There are several driving forces for the currents, which must be evaluated separately for proper current analysis and forecasting. The speed and direction of the resultant surface current (W) at a given point, time and depth below the surface (e.g. at a depth of 3 m) can be represented as the resultant of the following components;

(1) permanent flow (gradient current and/or 'characteristic current');
(2) wind current (lately often erroneously called the 'Ekman current');
(3) mass transport by waves;
(4) periodic portion of inertia current;
(5) periodic portion of tidal current;
(6) current caused by changes of atmospheric pressure and sea level; and
(7) the velocity and directional component, caused by influencing factors, such as changing depth of water, Coriolis force, and coast and current boundaries.

The direction and speed are considered to be specified for all these components.

Permanent flow is directly related to pressure gradients in the sea, which usually result from density differences of surface water columns between different locations (thermohaline circulation). The permanent flow is maintained by the great permanent and relatively constant wind systems (wind-driven circulation), which cause piling up of water masses and changes in sea level, resulting in pressure gradients. It is believed by some investigators that the permanent flow is in most areas an inertia function, due to prevailing winds, and is therefore frequently referred to as the characteristic current (Palmén 1930).

For any synoptic analysis and forecast of actual surface currents it is necessary to separate the relatively persistent, slowly changing permanent flow from other components of surface currents, which may vary hourly and daily and which should be predicted using the direct, synoptic influence of their driving forces.

Formulae have been derived for computation of currents from a pair of hydrographic stations, utilizing the difference in the dynamic heights at the stations. However, the following just criticisms have been raised concerning the application of these methods:

(1) The requirement that the stations in the section are quasi-synoptic is seldom fulfilled.
(2) The computation of dynamic height difference requires an assumption of a level or layer of no motion, which might exist but, at best, is difficult to assess correctly. This layer is generally assumed to be between 500 and 2000 m.

Saelen (1963) has made extensive comparisons of geostrophic, computed and measured currents in the Norwegian Sea, and found only a qualitative agreement, concluding that the concept of the circulation of surface and deep waters in the Norwegian Sea as a slow and regular drift does not hold, and currents at all levels vary rapidly over short periods of time.

The wind is one of the main driving forces of surface currents, and many current systems are mostly wind-driven. Wind is directly or indirectly one of the main causes of temperature fluctuations in the sea and thus also affects the permanent flow.

In numerous cases a simple linear relationship between the wind and current velocities has been found and/or used in the past. The current speed has been found to vary from 1.2% to 4% of surface wind speed. There is evidence that the factor decreases with increasing wind speed, and that fetch length and wind duration also have an effect on wind currents.

A simple relationship can be used to approximate wind drift from climatological data (monthly resultant wind speeds), a factor K times wind speed. The factor K varies between 0.014 and 0.025. The value of 0.02 has been found most appropriate for a rough estimation of the sum of average wind and wave current in the surface layer (at about 3 m depth).

The formula of Witting (1909), based on more than 5000 observations, has been generally found to represent best the relationship between the actual wind and actual current velocities. Current speed (cm s^{-1}) equals a factor K times the square root of wind speed (m s^{-1}). The factor K equals 3.8 if the mass transport by waves (wave current) is estimated separately. The mass transport by waves decreases rapidly exponentially with depth.

Recent investigations indicate that the wind current in offshore areas in middle latitudes deflects in general less than 20° to the right of the wind. If the wind is weak, other current components can prevail, and the current movement is erratic in relation to the wind. At lower wind velocities the angle of deflection is somewhat larger, about 25 to 30° in higher latitudes, but converges rapidly between wind speeds of 4 to 6 knots to about 20°.

A knowledge of tidal currents is of great importance for navigation in coastal waters, especially in inlets. Therefore predictions of tidal currents are made for such locations, usually using hydrodynamical numerical models. Tidal current analysis and prediction for offshore waters have been greatly neglected, and it has been generally assumed that tidal effects are weak over deep water. Measurements indicate, however, that tidal currents can have considerable magnitude even in mid-ocean. Some specific aspects of tidal currents are pointed out below:

(1) The tidal currents are usually rotary offshore and reversing in coastal waters. The tidal ellipses get narrower when one approaches the coast.
(2) In the northern hemisphere the sense of rotation is usually, but not always, dextral (clockwise).
(3) In general, the flooding current runs in the direction of the movement of the tidal wave (the direction can be ascertained from a co-tidal line chart).
(4) The rising tide is not everywhere synonymous with the flooding current, and falling tide does not necessarily mean ebbing current.
(5) High tide does not necessarily mean strong current, and a strong current may accompany a small tide. It is, therefore, always necessary to distinguish clearly between tide and tidal currents. The relations between the two are not simple and not the same in all locations.
(6) Semi-diurnal tide-producing forces cause proportional changes in tides and currents, but daily forces affect the current only half as much as the tides.
(7) The ebb usually lasts longer than the flood in coastal waters. How this can apply to rotary or nearly rotary currents offshore is not known.

(8) The topography of the coastal area influences the tidal currents considerably, causing eddy formation behind headlands and in bays.

When there is no apparent resultant force, except the deflecting force of the earth's rotation, acting upon a quantity of water in a given locality, the current is referred to as an inertia current. The primary force causing the inertia is usually the past prevailing and/or stronger winds in a given area, sometimes at some distance.

Available investigations on inertia currents show that they affect the speed of the permanent flow at the given locality by periodic fluctuation. At present no method exists for making a hindcast or forecast of these fluctuations. However, large and fast inertia eddies are often observed.

Theoretically, it can be expected that in the open ocean an atmospheric pressure change will cause a wave in the pycnocline rather than a true surface current. However, there must be some current and mass transport connected with this wave. No direct empirical evidence is available on the current component caused by the change of atmospheric pressure.

Internal waves, with tidal periods, may cause intermittent currents in the surface layers. Current speed and change of direction with depth may be related to internal and bottom friction. No exact model of change of current speed with depth exists. On the basis of theoretical considerations, the various components of surface currents differ considerably with depth. The pure tidal current should have uniform motion from the surface down, except very close to the bottom. In an inclined wind drift with friction, the depth–speed graph is a parabola. The depth change of mass transport by waves is exponential. The permanent flow can be considered to have uniform speed with depth in the mixed layer but a relatively large speed gradient at the pycnocline.

The currents near the coast can be classified into two systems;

(1) coastal current systems, consisting of relatively uniform drift, approximately parallel to the shore, composed of tidal, wind, and gradient currents; and
(2) near-shore current systems, consisting of coastward currents at the surface (wave current), a longshore current and seaward flow along the bottom and in the current rips.

The changing depth on the continental slope and shelf also influences current speed and direction. In general, the stronger currents are found along depth contours where the depth changes more suddenly, rather than following the configuration of the coastline. Usually the strong permanent currents flow along the continental slope.

Currents flowing along the coast with a slight offshore angle cause upwelling.

Eddying and upwelling are also caused by steady currents along the coast if the direction of the coast changes. Islands in a steady current cause eddies on the leeward side and modify the current pattern at a considerable distance from the coast. In semi-closed bays, there is usually a conformal current along the coast, very similar to the currents observed in lakes. The tidal currents are usually dominant on the continental shelf. Upon these tidal currents there is a wind current component and near the estuaries a haline gradient component caused by run-off.

The hydrographic and especially the dynamic conditions at major current boundaries are usually rather complicated. The types of current boundaries are;

(1) dynamic (divergences or convergences of current systems);
(2) topographic (caused by the topography of the bottom or coast); and
(3) combined eddy systems.

The most usual locations of the current boundaries are near meteorological fronts, on continental slopes, and around islands, capes and banks (local boundaries).

The boundaries are frequently marked by current rips, accumulation of flotsam (surface slicks), modified waves, roaring noises and fog. Usually there is also a change of water colour at the boundaries. The surface temperature gradient on the boundaries can be 0.5° to 2° per 10 nautical miles but may be even larger for major currents.

The current boundaries, especially convergences, are associated with meanderings and eddies which cause sinking and/or upwelling of deeper water. The greater the speed difference on both sides of the boundary, and also the more irregular the bathymetry of the bottom, the more extensive is the meandering. Many of the large-scale meanderers are stationary, owing to the topography of the continental slope. However, their intensity seems to fluctuate, depending on changes in driving forces.

There are also great eddy systems in the ocean which are caused by winds (cyclones and anticyclones). In general, the centres of cyclonic eddies (anticlockwise) are cold, and they are associated with slow upwelling in the centre. The centres of anticyclonic eddies are warm and sinking occurs very slowly in the centre. If a cyclonic eddy is cut off from a cold water mass into a warmer one, slow sinking also takes place in this eddy.

The effects of currents on fishing operations and their effects on availability of fish are described in Chapter 4 section 4. The climatological effect of changing current systems, brought about by climatological changes in surface winds, is described in Chapter 3 section 3. The importance of these climatological changes pertaining to currents has been underlined by Flohn & Fantechi (1984) as follows:

'The variations of the wind circulation appear to be accompanied by variations in the ocean surface currents also, affecting the course and strength of the main flow of Gulf Stream–North Atlantic Drift water across the North Atlantic and the fluctuating position attained by the boundary between this warm saline water and the polar water flowing south from the area northeast Greenland-Spitsbergen.

These variations of winds and ocean currents are clearly of importance in connection with many aspects of the climate of Europe. They are intimately connected with the variations of prevailing temperature and rainfall and the incidence of occasional long spells of one or another kind of extreme weather. They also affect the movements of fish stocks in the sea and hence the fisheries. More detailed investigations are needed, including the large-scale spatial correlations (teleconnections) between climatic anomalies in western and central Europe and other regions of the world, having in mind the economic role of such anomalies and their interdependence.'

The main permanent current systems are depicted in monthly and seasonal surface current charts, available in numerous atlases. The currents in these charts are deduced from ships' drifts. Most of these surface currents are wind-driven, either by local winds or by distant wind fields. Climatic changes in the oceans pertain mostly to relatively small changes in these current systems.

2.5 Ocean thermal structure and sea ice and their spatial and temporal changes

Although ocean thermal structure implies three-dimensional temperature distribution in the ocean, it will be treated below in two aspects, sea surface temperature and temperature structure in the upper layers of the ocean down to about 200 m.

Heat exchanges between the atmosphere take place through the sea surface, whereby the sea surface temperature (SST) plays a major role in these processes. SST is the most observed parameter in the sea, and is also a good indicator of various processes in the surface layer which have occurred in the past.

Observations of thermal structure below the surface are relatively sparse and subsurface temperature climatology rather than synoptic analysis must be used in many applications. The most important aspect of this subsurface thermal structure in respect to fish and fisheries is mixed layer depth (MLD, the depth of the top of the seasonal thermocline). This depth can be computed from surface driving forces, using climatology also.

The sea-surface temperature (SST) is herein defined as the average temperature of the surface mixed (turbulent) layer at a defined position and time. The isothermal surface layer may be 1 to 3 m thick during heating season in calm weather to about 200 m during the winter cooling season. It usually extends from the surface to the depth where the mixing by the strongest wave motion in the present or recent past (a few days to 2 weeks) has caused no appreciable mixing. During the cooling season the surface turbulent mixing nearly always reaches to the top of the thermocline owing to convective stirring.

The SST is one of the most easily measured environmental properties in the sea. It is the only truly oceanographic element which is satisfactorily observed and reported on a synoptic schedule.

The long-term variations of SST have been pointed out already by Helland-Hansen & Nansen (1920) who demonstrated how these variations affect the variations of climate in western Europe and showed the relations of SST to a number of other factors in the sea and in the atmosphere.

Synoptic analyses and forecasts of SST are finding use in synoptic meteorology through the inclusion of heat exchange effects into meteorological forecasting models. Furthermore, SST is required in the computation of changes of surface air properties over the ocean, in frontal and heat exchange forecasts, in wave forecasting and in prediction of visibility and probability of fog. SST analysis is used in the computation of surface currents and in the verification of current forecasts (Hubert & Laevastu 1970). The SST is used also in forecasting icing conditions of ships. It is used further as one of the basic parameters for the prediction of ice formation in high latitudes such as the Baltic Sea and the Arctic.

At times the temperature itself might not be the direct affecting factor we are looking for, but it might be used as indicating other changes and conditions in the sea. Examples of indirect uses are the estimation of upwelling intensities and the computation of current and surface water type boundaries. Correlations between temperature and the behaviour and occurrence of fish have been sought and found. An extensive summary of this subject is given by Laevastu & Hayes (1981).

Analyses and forecasts of SST, on both the synoptic and climatological scales, can be used in fisheries for the prediction of the delay and/or advancement of spawning times and displacements of spawning grounds and as one of the factors in the prediction of the survival of larvae. The temperature boundaries, i.e. the convergences and divergences of currents, are often important fishing grounds for pelagic fish, and the forecasts of their movements are sought.

The factors which determine and modify the surface temperature distribution and subsurface thermal structure may be divided into five groups,

(1) heat exchange,
(2) convective mixing,
(3) mechanical mixing,

(4) advection, and
(5) diverse dynamic factors usually of a local nature.

The heat exchange group of effects includes several semi-independent components. First, insolation is computed from solar altitude, length of day and cloud cover. In this component, a good determination of the type and amount of cloud is the most difficult problem. Secondly, the transfer of latent heat by evaporation and condensation is computed from analyses of environmental parameters at the interface (water vapour pressure difference and wind speed). The third component involves the exchange of sensible heat, the fourth concerns reflected heat (a function of the albedo of the sea surface), and the fifth is the effective long-wave back radiation.

Mixing distributes the heat vertically in the oceans and to a large extent determines the temperature structure with depth. In areas of heat loss, the stirring is due to convective mixing, by which surface water becomes heavier owing to cooling and evaporation and therefore sinks. In areas of positive heat exchange, the depth of the thermocline and the gradient within it are determined by forced or mechanical mixing through wave action. This process also affects the SST by mixing deeper, colder water from below the thermocline into the surface layers.

Local processes of heat exchange and mixing are modified by advection due to currents and the accompanying effects of divergence or convergence and to some extent by horizontal mixing. Divergence causes upwelling and a thinning of the surface layer, while convergence brings about a piling up of surface waters and deepening of the mixed layer. In most medium and high latitude areas the advectional changes of SST are larger than other change components.

Included in the fifth group of factors affecting SST and the subsurface thermal structure are a number of diverse factors which have local effects. Some of these factors are transfer of heat by rain and melting snow, transport of fresh water from estuaries, and mixing by currents (e.g., in coastal areas by strong tidal currents).

This group of factors also affects the SST distribution and changes as well as the subsurface thermal structure in a given fixed location through eddies in various scales, mainly inertia currents, and the movement and meandering of current boundaries.

Numerous charts of long-term monthly average SST have been published. However, these charts differ somewhat from each other, depending predominantly on the period of data and averaging techniques used. The average SST shows considerable year-to-year and longer-period changes, which are different in different locations. For any computation of SST anomalies there is a need for one extra long-term average which would serve as a basis for comparison. Hemispheric climatology of an acceptable accuracy is available at present in the atlases prepared by Robinson (1976) and Robinson *et al.* (1979).

Several attempts have been made in the past to predict long-term SST changes. Long-term analyses and predictions of SST should really be called anomaly investigations. The greatest variations of surface temperature in a given locality are frequently caused by advection due to wind-driven currents. As long-term wind predictions are not possible, so neither are SST anomaly predictions.

An example of a monthly SST anomaly is given in Fig. 2.14. The anomalies are largest in the regions of main oceanic current systems and along the east coasts of the continents where the annual range of SST is largest (Fig. 2.15). Thus, often the SST anomalies indicate a change of timing of annual fluctuation of temperature. The SST anomalies over the ocean have a comparative space scale of about one-quarter of the main pressure and wind systems (e.g., the Icelandic Low).

A typical seasonal change of thermal structure below the surface in medium and higher latitudes is shown in Fig. 2.16. A number of typical temperature profiles with key features, such as thermocline, are shown in Fig. 2.17. In some cases the thermocline is so extended (B, Fig. 2.17) that its lower boundary has not been determined. This type can be derived from the continuous density model (E), which has been modified by wave action and by currents into the epithermocline (C). If warmer water flows over homothermal water (G), the structure (F) may result; and if it is heated in calm water, type (D) is observed. If a homothermal water is flowed over by less saline or colder water, a temperature inversion (H) occurs. In waters which are stratified by salinity and where effective mixing between the various strata is prevented, several thermoclines (I) may be present.

The thickness of the mixed layer and its aperiodic and periodic undulations are determined mainly by the following factors;

(1) Mixing by wave action,
(2) convective stirring caused by cooling and excessive evaporation at the surface,
(3) divergence and convergence of the currents (upwelling and anstau),
(4) large-scale eddies, and
(5) internal waves and tides.

Short-term forecasting of mixed layer depth (MLD) is based on a knowledge of wind mixing, heating and cooling and surface current convergence or divergence. During the heating season mixing is due primarily to wind waves. The relationship between significant wave height (in the last 24 h) can be used for MLD prediction during the heating season; MLD equals ten times the significant wave height.

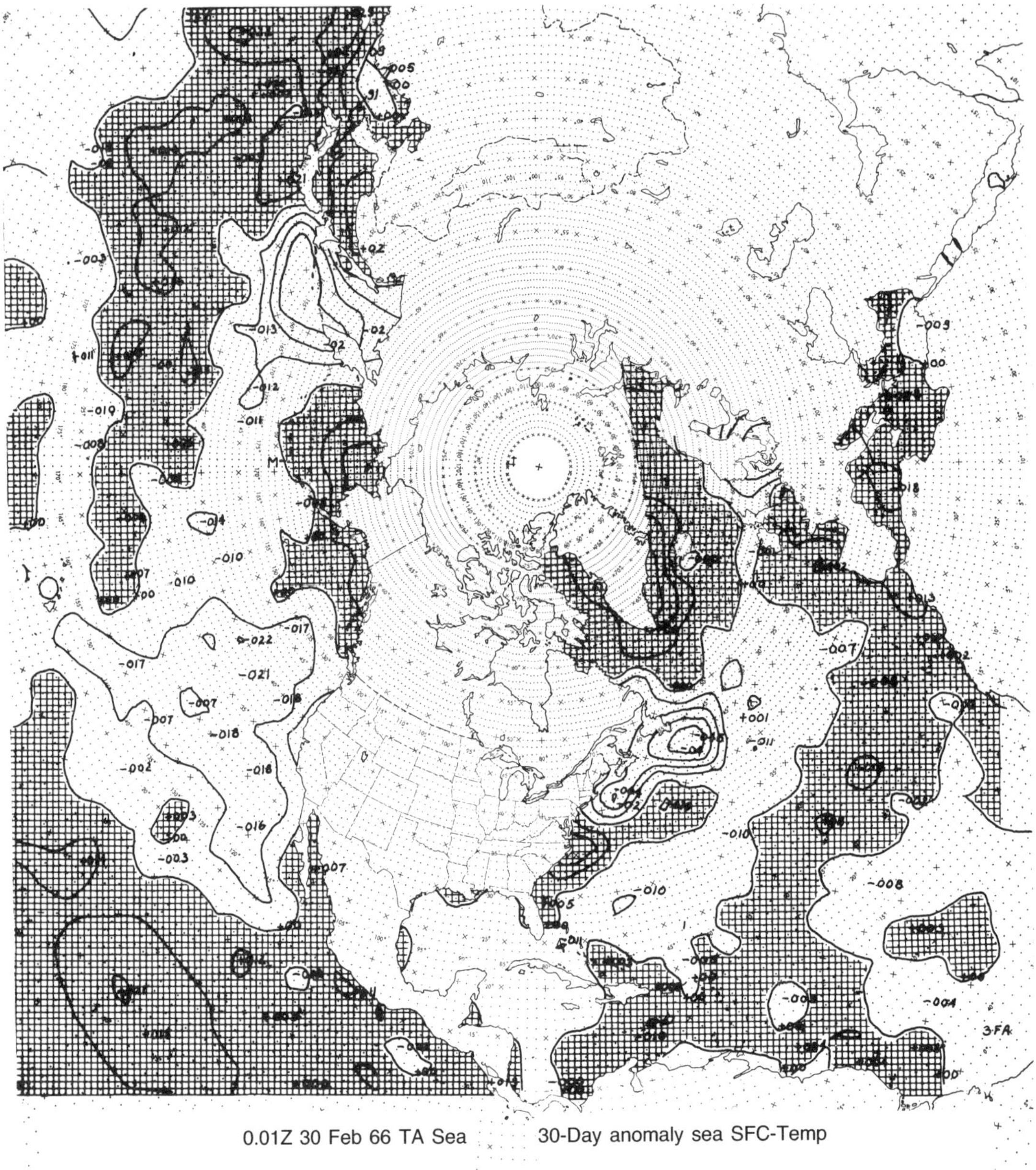

Fig. 2.14 Monthly sea surface temperature anomaly, February 1966.

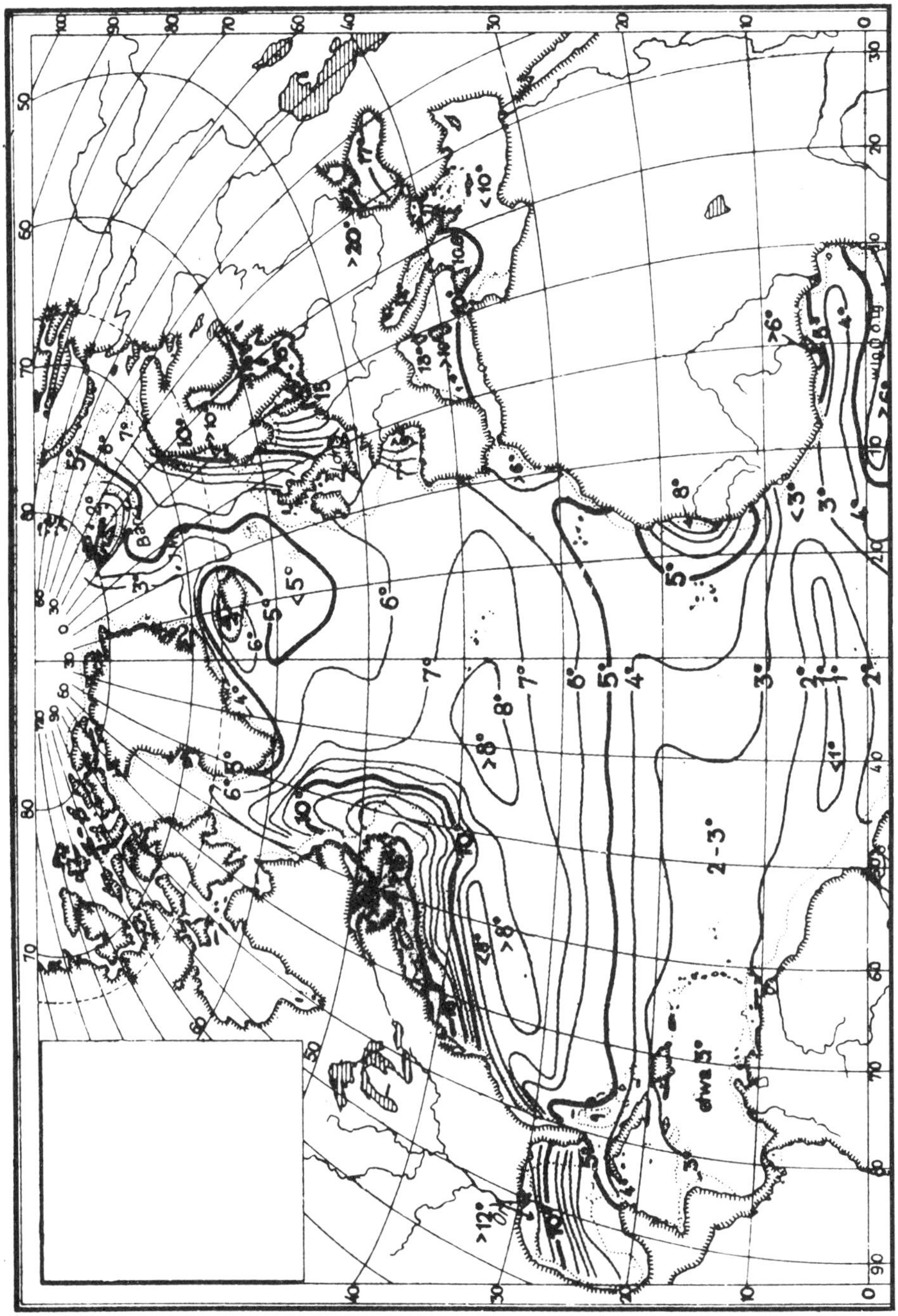

Fig. 2.15 Annual range of sea surface temperature in the North Atlantic (Schott 1944).

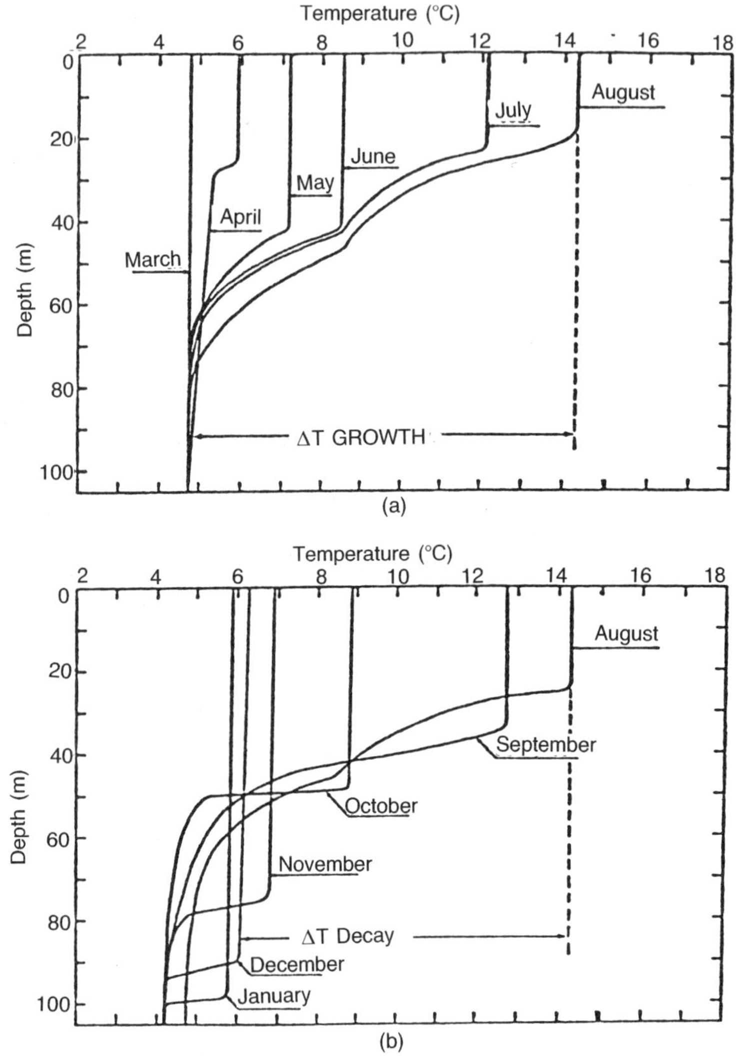

Fig. 2.16 Time variation of mixed layer thermal structure at Ocean Weather Station *P*. (a) March–August, (b) August–January (Tabata 1983a).

During the cooling season convection becomes the dominant mixing process. The potential layer becomes isothermal (mixed), and the layer depth sinks slowly and regularly (about 20 m per month). Wind mixing and convective mixing are independent and additive processes. Convection always creates a truly mixed (isothermal) layer, which is deeper than a simple wind-mixed layer. When it occurs, convection determines the character and depth of the layer, and wind mixing can be disregarded.

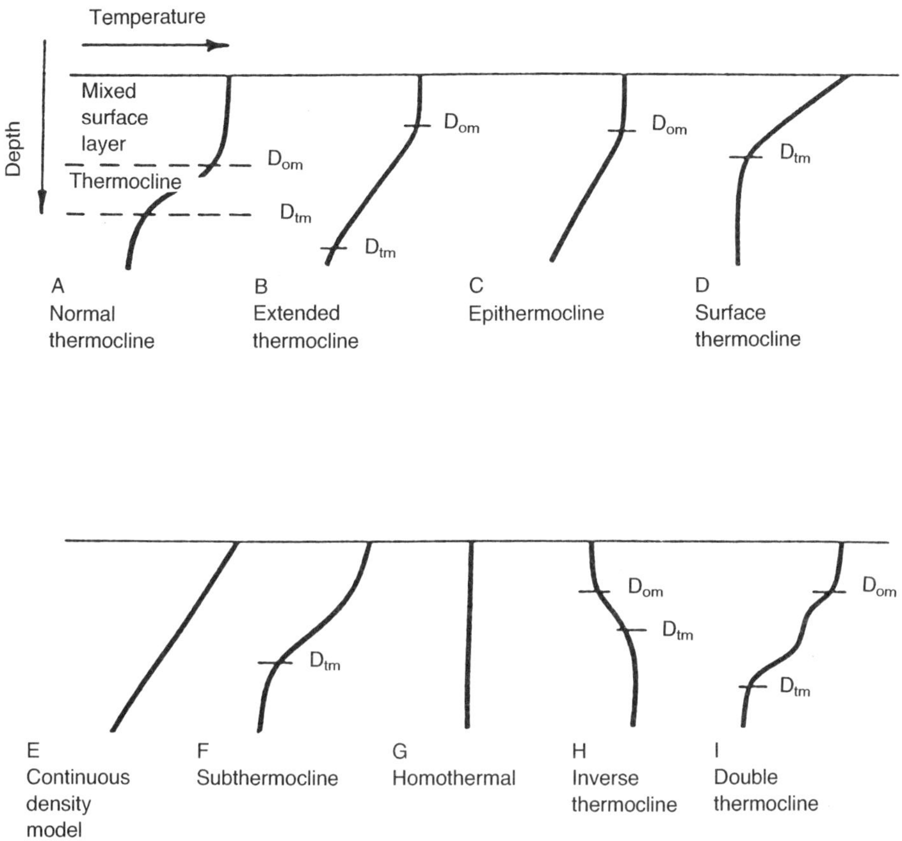

Fig. 2.17 Types of temperature structures of the upper layer of the sea.

Considerable vertical undulations are observed in the depth of the thermocline. These undulations have a number of causes, swell at the surface, internal waves caused by a variety of forces, tides, atmospheric pressure change, etc. There are ocean areas where the undulations are small and other areas where they are, in general, large. The mean range of fluctuations is between 3 and 8 m.

SST anomalies indicate climatic changes in the surface layers of the sea. Some time series SST anomalies in a few locations are given below as examples of their magnitudes and nature.

Tabata (1975) summarized the monthly SST anomalies at Weather Ship *Papa*, in the middle of Gulf of Alaska, and at the closest coastal station (Pine Island) from 1950 to 1975 (Fig. 2.18). The significant fact is that the coastal SST anomalies are quite different from those in offshore areas. No periodicities of anomalies are apparent, nor are there any real and significant

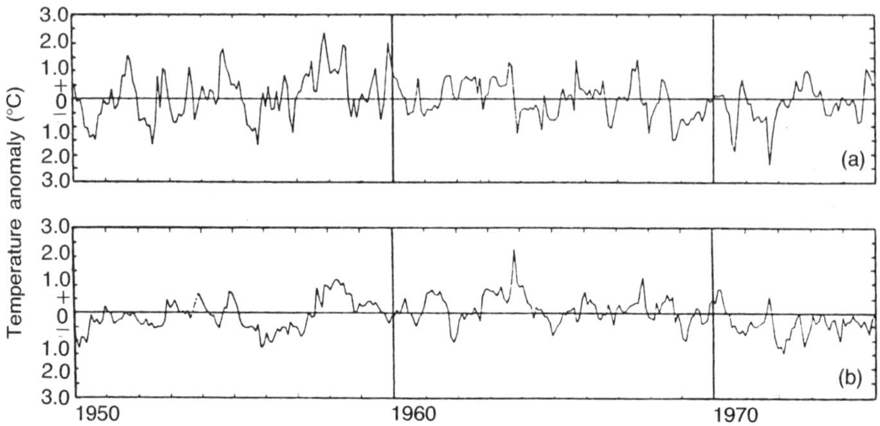

Fig. 2.18 Anomaly of monthly sea surface temperature (°C) based on 22 years' means, 1950–1971 (Tabata 1975). (a) Station *P* (50°N, 145°W); (b) Pine Island, Northern tip of Vancouver Island.

long-term trends in these anomalies. Tabata (1983b) also reviewed a number of past studies on the possible effects of the North Pacific ocean on the climate. He found that most of the studies were conceptual in nature and, although some evidence was provided to back some of the hypotheses, none has been adequately tested.

Temperature changes along the Greenland coast are of particular interest, as the region is the northern (colder) boundary region of the distribution of commercially important fish, especially cod, and large-scale changes in cod fisheries have occurred in this area. Herman (1967) has summarized the SST fluctuations in two areas around Greenland from the 1870s to 1962 (Fig. 2.19), which show the significant increase of temperature in the mid-1920s. There has also been a decrease in the difference between winter and summer air temperatures (i.e. warmer winters) starting in about 1905. Blindheim (1967) has shown that large changes can occur in the Irminger part of West Greenland current, and thus the temperature changes in the Greenland area might be caused mainly by advection.

Large temperature anomalies in the ocean are relatively slow to change, and may last throughout the winter or summer when once established in the autumn or spring. The magnitude of these anomalies is usually in the order of 1 to 2°C, seldom rising above 3°C. It has been postulated and shown in the past that these SST anomalies can cause atmospheric anomalies downstream through feedback processes (Lamb & Ratcliffe 1969).

Ratcliffe & Murray (1970) have used the feedback relationships between the Atlantic sea surface temperatures upstream (east of Newfoundland) on the one hand and monthly atmospheric circulation anomalies on the other

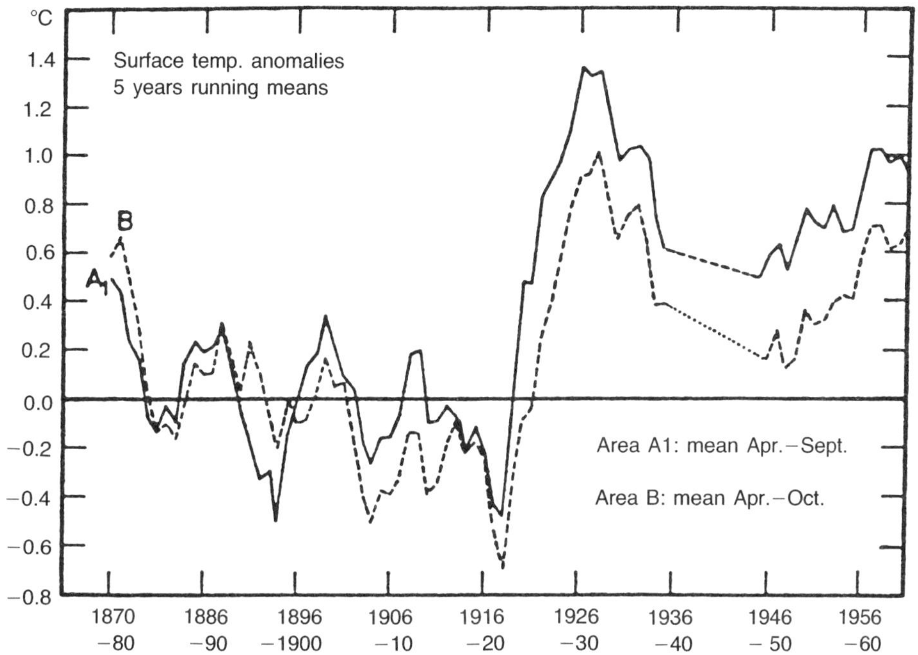

Fig. 2.19 Sea surface temperature anomalies for area A1 (West Greenland) and area B (South Greenland), 5 years running means (Herman 1967).

hand to attempt long-range predictions of weather over the British Isles and in western Europe. The key area in the Atlantic is a wide area south of Newfoundland; colder than usual ocean surfaces in this area were shown to be associated with blocked atmospheric patterns the following month over northern and western Europe while a warmer than usual ocean in the same general area favoured more progressive synoptic types to follow.

The depressions normally travel east or northeast near Newfoundland and generally reach their maximum intensity well downstream, the precise track depending greatly on the general tropospheric flow pattern. The extra water vapour transferred from the ocean to the atmosphere may be carried many hundreds of miles downwind in the upper flow before detectable heat is released by the condensation process. These processes operate on a time scale of a few days but recurrences over the period of a month or so appear to result in statistically favoured areas downstream, depending on the season, where the complex processes will be manifest and marked by centres of negative pressure anomaly. However, with lower than usual sea temperature in the energy source region south of Newfoundland it is likely that downstream developments will be different; in particular, cyclonic and anticyclonic tendencies are likely to be located in different geographical areas compared with the warm sea.

In addition to the heat exchange processes already suggested, a strong gradient of sea temperature anomaly may act as a weak steering control for atmospheric systems; cold air tends to be more conservative over cold sea and warm air more conservative over warm sea, so that a strong sea temperature gradient will tend to maintain any baroclinic zone in the atmosphere which is approximately parallel to it.

The subjective long-range (monthly) weather predictions by Ratcliffe and Murray above-described are among the very few where some cause and effect bases have been established and where use is made of sea surface temperature anomalies.

2.6 Chemistry and biology

The chemistry of sea water which might affect fish is little influenced by weather and by climatic changes. Two aspects should, however, be considered, salinity and basic nutrient salts such as phosphates and nitrates, which can be limiting factors in basic organic production (phytoplankton production) in the sea.

Salinity in coastal waters is affected by run-off, especially from larger rivers, and its distribution with currents and mixing. Thus regional changes of precipitation and man-made changes in river run-off (e.g. river diversion and/or extensive use of water for irrigation) can affect salinity in estuaries and in their vicinity, which in turn might affect local fish ecosystems. On an oceanic scale the changes of salinity in surface layers are small and are affected by changes in evaporation-precipitation balance, ice melting and freezing, and in surface currents (advective anomalies). However, the salinity of the surface layer is a relatively conservative property. If it is once changed in a given area and time, it can persist for a number of years while the water is advected by surface currents to other areas (e.g. 'The Great Salinity Anomaly' in the North Atlantic in the 1970s).

The limiting nutrient salts in the surface layers of the sea, mainly phosphates and nitrates, can be affected by run-off which brings organic domestic and agricultural pollution from land areas. This addition of nutrients, called auxotrophication or eutrophication in limnology, might increase phytoplankton and related zooplankton and fish production, as occurs in the Baltic Sea. The supply of nutrients to the surface layer is also determined by vertical mixing, whereby the winter turnover on continental shelves in medium and high latitudes plays a predominant role. Vertical mixing is effected by winds and surface cooling, both of which are elements of weather and climate.

Some of the biological production, contents and processes in the sea may be affected by weather and climate fluctuations, which will in turn affect the fish ecosystem mainly through fish food and its availability. The basic organic production (or phytoplankton production) is affected by;

(1) the availability of limiting nutrients,
(2) the amount of light (sunshine),
(3) the depth of mixed layer,
(4) temperature, and
(5) transport by currents.

Production of zooplankton, which is the main food for most juveniles and small pelagic fish, is affected by the availability of phytoplankton, temperature, and the transport of eggs and nauplii. It could be noted that the local zooplankton production is in many areas dependent on brood from other, upcurrent areas. Consequently zooplankton production in a given region can be influenced by climatic fluctuations, especially changes in advection (currents) and in phytoplankton production, which in turn is influenced by climatic fluctuations.

Ichthyoplankton (fish eggs and larvae) is affected by the availability of phyto- and mainly zooplankton as food, temperature, transport, and especially the numbers of predators present. Benthos on the continental shelf is an important food source for demersal fish. Its abundance can be affected by temperature and the availability of detritus, through organic pollutants, from phytoplankton or from demersal seaweeds).

Intensive studies by ICES (The International Council for the Exploration of the Sea) in the 1970s failed to find any increase in phosphates in the North Sea which might have accounted for increased basic organic production. However, the studies of the Helsinki Commission (the Baltic Marine Environment Protection Commission) found in the Gulf of Finland an annual increase of total phosphorus of 2 to 3% due to eutrophication, but no significant changes of phosphates in the surface layers (Kahma & Voipio 1989). The available data did not permit the analysis of the change of duration of spring blooming in the Gulf of Finland, which might have become slightly longer, allowing higher annual organic production.

Colebrook (1979) analysed the seasonal cycles of phytoplankton and copepods in the North Atlantic Ocean and North Sea, using continuous plankton records taken over a number of years. He found a clear relationship between the surface temperature and the timing of the spring increase of phytoplankton and the amplitude of the seasonal variation. Spring increases were found to occur in the North Sea, associated with transient periods of vertical stability, resulting in a slower rate of phytoplankton increase. According to Glover *et al.* (1974) the analysis of the same plankton records from 1948 to 72 showed that the total number of copepods decreased, the duration of abundance decreased, and the Atlantic Ocean surface water type extended further north, with a reversal in the last years of the study (Fig. 2.20).

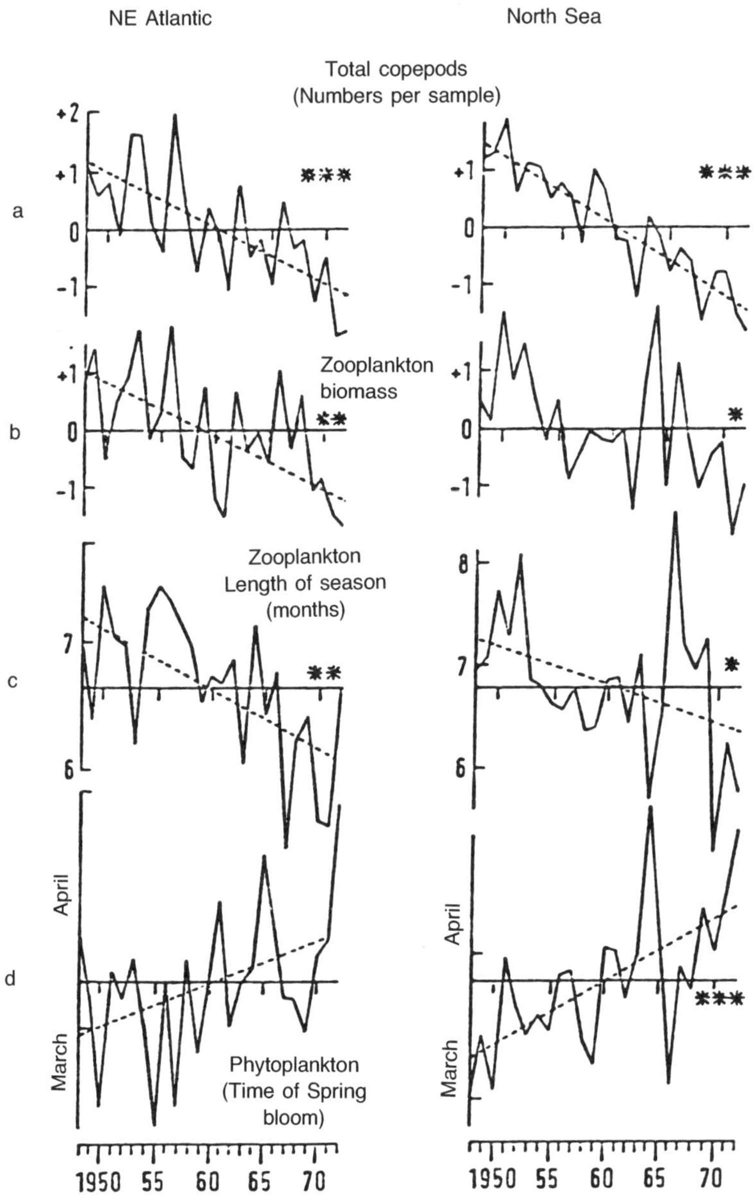

Fig. 2.20 Changes in the North Sea and NE Atlantic between 1948 and 1972 as shown by the plankton recorder network; (a) numbers of copepods; (b) zooplankton biomass; (c) length of zooplankton season; (d) timing of spring bloom (Glover *et al.* 1971).

The amount of planktonic fish food available in a given area has been considered by Japanese scientists as an indication of the possible availability of pelagic fish (Burbank & Douglass 1969). Higher amounts of zooplankton are found at current boundaries. This aggregation can occur either dynamically (Laevastu & Hela 1970) or through mixing of nutrient-poor oceanic water with richer coastal water. This condition was described by Fraser (1961), who found that herring do not appear in water of Atlantic origin flowing into the North Sea. After this water mixes with North Sea water a richer zooplankton develops. Whether the absence of herring in Atlantic water is caused by herring leaving this water because of lack of food, distaste of the water itself, or because of the presence of salps or other oceanic species, is unknown. The relationship between food abundance and pelagic fish is a general but not a simple one. For example, salps and jellyfish can develop rapidly and graze down zooplankton, and jellyfish might act also as fish repellents.

In earlier years it was assumed that starvation might be one of the main causes of extensive larval mortalities. However, lately it has been shown that this might not be true. For example Munk & Kiorboe (1984) showed that herring larvae can feed satisfactorily in relatively low food densities. Recruitment of commercial species and the fish ecosystem at large is mainly controlled by predation on ichthyoplankton and juveniles (for a summary see Bax & Laevastu 1990). Predation pressure is especially heavy on small pelagic fish larvae such as herring and sprat (Flintegaard 1981) and pelagic larvae of demersal and semidemersal species such as plaice and cod (Daan *et al.* 1985).

Food preference in a given species can vary somewhat from stock to stock (e.g. between coastal cod and migrating cod in Norway, Sunnanå 1984). It can also vary from area to area and season to season, depending on absolute abundance of the prey types and on seasonal redistribution of the species (Daan 1981). Hempel (1973) concluded that plankton or benthos biomass in general is normally not a limiting factor in size and productivity of population of fish. There are two cases where plankton production and its changes due to man's action and climatic change might have caused changes in fish ecosystem and its productivity: Baltic Sea and Atlanto-Scandian herring (see Chapter 6 section 2 and Chapter 7 section 2).

2.7 Coastal weather and the ocean

The coastal ocean, i.e. the continental shelf near the coast from a few miles to about ten miles offshore, the bays and smaller semi-closed seas, was in the past the main site of fishing grounds and is still important to local fisheries and mariculture. It is the part of the sea most affected by man (e.g. by pollution) and might also be more sensitive to climatic fluctuations than the open ocean. From the point of view of weather and climatic fluctuations, the following conditions and processes in coastal areas are briefly considered in this sub-section; temperature changes, local winds and waves, sea level changes, run-off and ice. The pollution problems are reviewed in Chapter 6.

Water temperature has been observed in many locations along the coast for a considerable length of time. Blindheim *et al.* (1981) have reviewed the long-term temperature trends in Norwegian coastal waters. Although the individual monthly mean temperatures show considerable differences from station to station (Figs 2.21, 2.22), 5-year running means show fair agreement between observations at all stations in summer and winter and display fluctuations with a duration of between 5 and 20 years. These observations suggest that the fluctuations of temperature along the Norwegian coast may

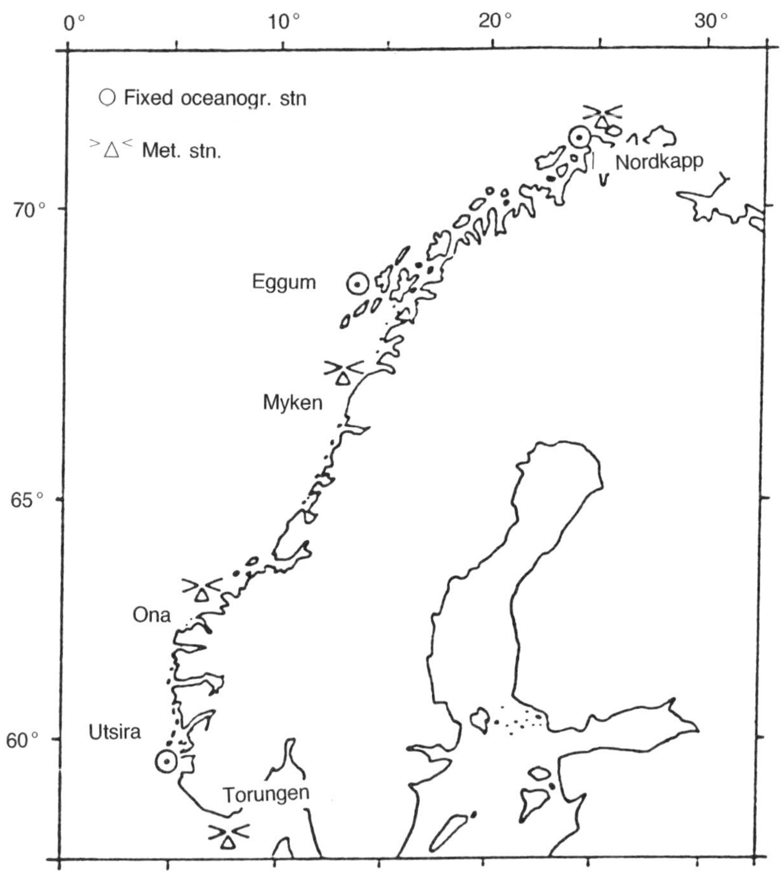

Fig. 2.21 Position of observation of sea surface temperature (Blindheim *et al.* 1981).

be of an advective nature. A long-term warming trend, which is more conspicuous in summer than in winter, started at the turn of the century. This temperature trend along the Norwegian coast is, however, different from that at Greenland (Fig. 2.19).

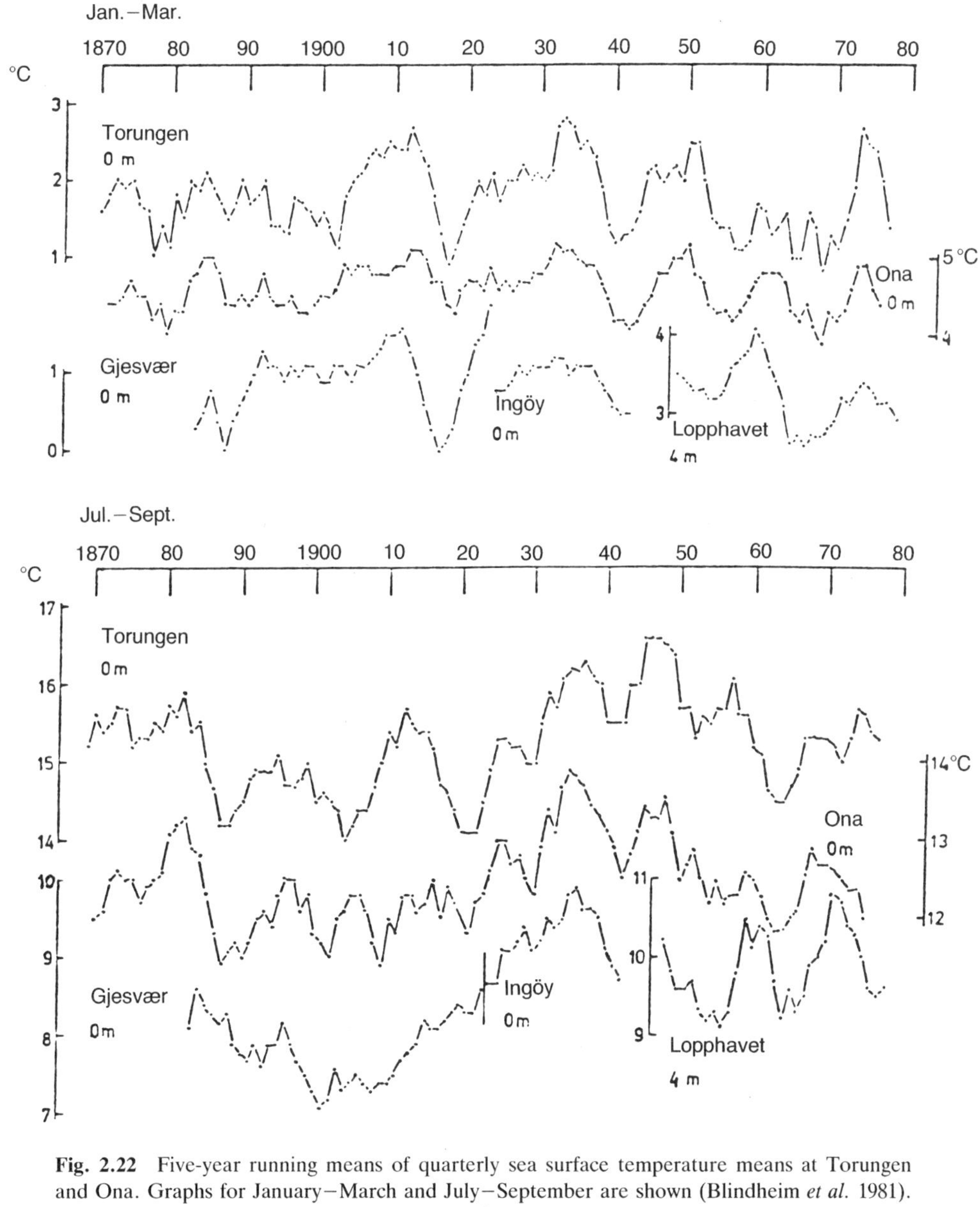

Fig. 2.22 Five-year running means of quarterly sea surface temperature means at Torungen and Ona. Graphs for January–March and July–September are shown (Blindheim *et al.* 1981).

Rose & Leggett (1988) found near-shore sea temperatures and currents on the north shore of the Gulf of St Lawrence to be linked to wind-forced upwelling and downwelling. Near-shore temperature fluctuations caused by upwelling are pronounced in major upwelling regions (Arntz & Valdivia 1985). Off low coasts the coastal winds are determined by the passing and/or prevailing surface meteorological systems. In some areas and seasons these winds may be modified by sea and land breezes.

The wave action in shallow water reaches to the bottom, especially during storms. Many fish species leave shallow water during storms, for reasons which are not clear. It is assumed that the particle movement by waves might affect the well-being of the fish and the bottom sediment whirled up by waves might get into the gills. Beds of sessile shellfish might get muddied up during the storms. Thus the variable frequency and force of the storms in coastal areas affect not only fishing but also the well-being of fish ecosystems. Coastal fisheries with small vessels are also affected by the frequency of storms.

Changes of mean sea level along the coasts consist of yearly variations and secular (non-periodic) trends. The greatest part of the yearly variation of sea level is due to variation of wind and its effects, such as wave set-up, net wave transport and wave-induced currents. Furthermore, atmospheric pressure effects are usually concomitant with wind effects; the sea acts as an inverted barometer whereby 1 mb change corresponds to 1 cm change of sea level. The wind and atmospheric pressure changes are usually connected with shifts of mean pressure systems. Therefore the annual sea level changes can be in phase over some distances (Fig. 2.23). Sea level changes are also affected by temperature anomalies (steric sea level) and by river discharges. Meade & Emery (1971) found that river discharges contribute 7 to 20% of the annual sea level variations.

The secular changes of sea level can be connected to secular changes of wind systems and to isostatic (eustatic) movement of the land. These changes can vary from station to station (Fig. 2.23) and from one region to another. The best known and persistent isostatic movements are in Scandinavia and Finland, where the annual lift-up might reach 1 cm per year.

The variation of sea ice from winter to winter in coastal areas can reflect climatic changes and would affect fisheries and to a minor degree fish stocks. Observations and forecasting of ice have long been practised in high-latitude countries, especially in the Baltic area. The occurrence and extent of ice in semi-closed seas, such as the Baltic, can vary considerably from year to year, without any recognizable periodicity (Fig. 2.24). However, there is a secular trend in the length of the ice season (a decrease) (Fig. 2.24), the rate of decrease having been highest between 1860 and 1900. The maximum extent of ice in the Baltic is usually related to winter air temperature.

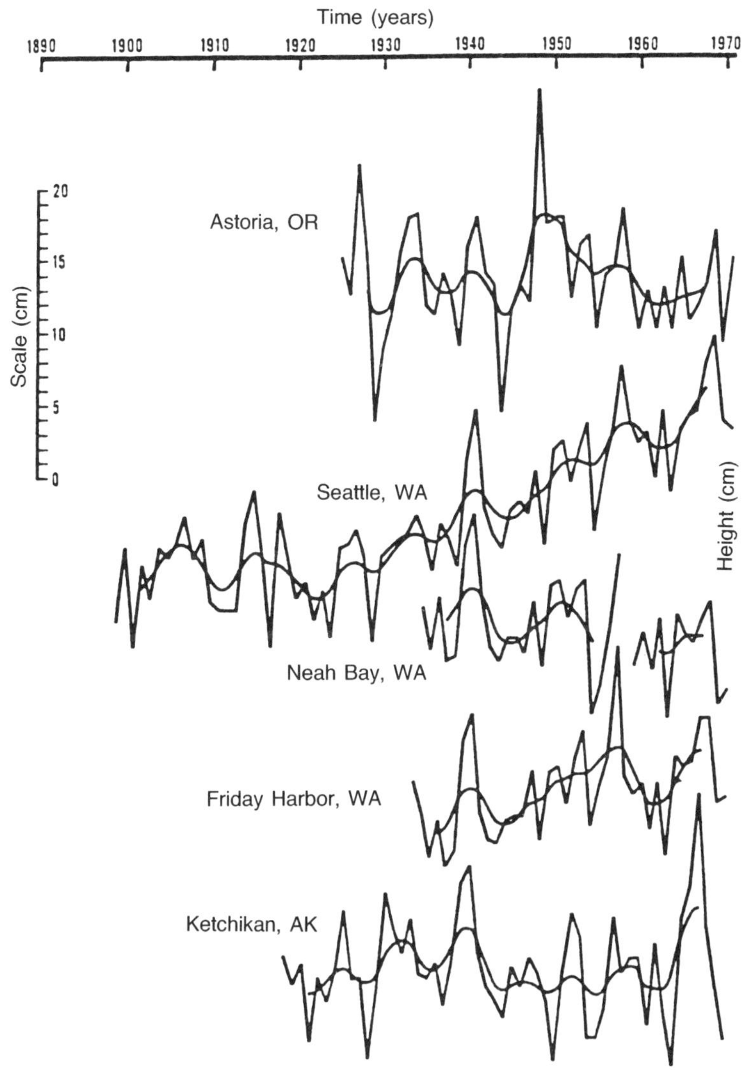

Fig. 2.23 Change in sea level with respect to adjacent land for stations from Oregon to Alaska (Hicks 1973).

The run-off will bring nutrients from rivers to estuaries and coastal areas. However, when this nutrient supply is integrated over a larger sea area its effect is small. Elmgren (1984) estimated that in the Baltic the pelagic primary production is 134 g carbon per m^2 per year, with river input of 13 g C, or about 10% of the carbon used in organic production; direct waste discharge of organic matter (2 g C) is only a minor energy input. An

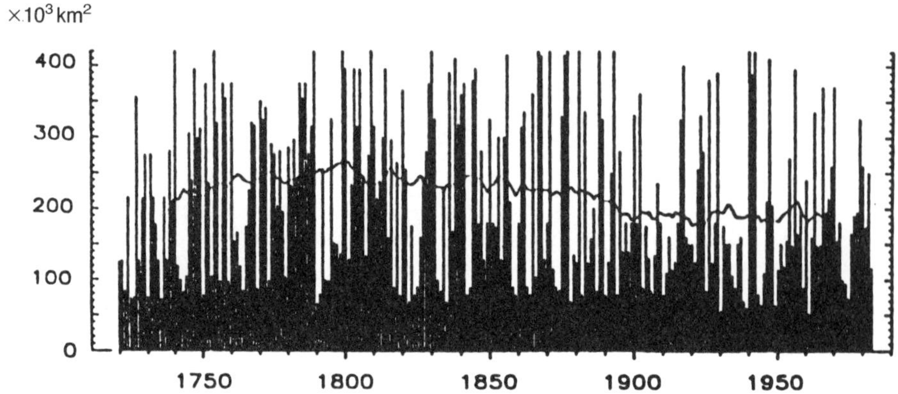

Fig. 2.24 The maximum extent of ice in the Baltic Sea 1720–1983. Yearly values and 33-year floating averages (Makkonen *et al.* 1984).

estimated 5 g C per m^2 per year is consumed by fish, its annual yield being only 0.2 g C per m^2 per year.

Coastal areas are important spawning and nursery areas for many species which are normally distributed offshore. Thus man-made and climatic effects in coastal areas might affect the stocks of these species. This aspect is reviewed further in Chapter 4 section 3 and Chapter 6.

Chapter 3
Marine Climates, Hydroclimes, and Their Relationships to Marine Resources

Marine climates refer to atmospheric climatic conditions and processes which determine and change them over the oceans. From the fisheries and marine biological points of view we are, however, mostly concerned with the aquatic environment of fish and its changes. Thus we call the aquatic climate the hydroclime. The main driving forces which determine and change the hydroclime conditions are atmospheric forces and processes — i.e. the marine weather and its integrated effect, the climate.

Within the general hydroclime there are a few conditions and processes of primary concern to fisheries, which need some definition by subject and by scope of concern. This is attempted in the introduction of this chapter, starting with the definition of space and time scales of the hydroclime. Vertically the hydroclime regime extends from the surface to the bottom of a permanent thermocline or, where there is no thermocline, to about 200 to 500 m depth. In deeper water the changes of properties with time are very small and slow, and an insignificant part of fish ecosystems reaches these depths.

The horizontal scales of hydroclime regions are considerably smaller than atmospheric climatic regions (zones). The hydroclime regions are often determined by the configuration of the coast and bathymetry of the basins (e.g. North Sea, Bering Sea) and/or by surface current features (e.g. Irminger Sea, Norwegian Sea). A division of the world's oceans into hydroclime regions or natural regions is given in Fig. 3.1. The division is in many aspects similar to the corresponding divisions of Schott (1935, 1944) and Dietrich & Kalle (1957). It is possible to characterize and describe these regions by their prevailing oceanographic, meteorological and biological conditions. That is to say, they must be natural rather than purely geographic or political regions. The main characteristics of the natural regions in Fig. 3.1 are given in Table 3.1.

Some ocean areas have been studied much more intensively than others and a great deal of information is available regarding them. Furthermore, there is considerable similarity between some regions. Therefore, by dividing the ocean into natural regions, one can apply information obtained from one well-known natural region to another similar natural region where the data are more scarce. In Table 3.2 similar regions are grouped together.

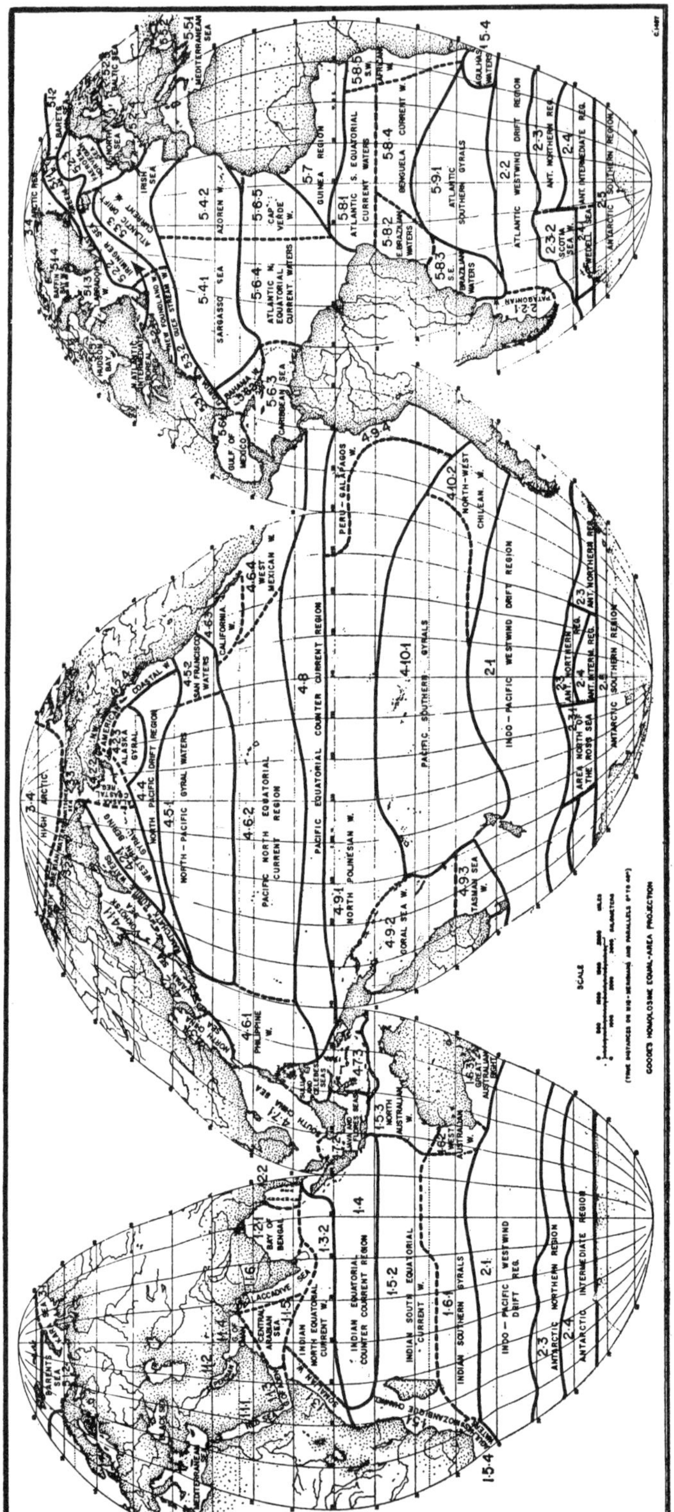

—— Boundaries of the regions
-------- Boundaries of the surface waters

Fig. 3.1 Natural regions of the oceans.

Table 3.1 Characteristics of natural regions in the oceans

No.	Names of the Regions and Subregions	Winds: Prevailing Direction (from)	Winds: Resultant Force (Beaufort)	Winds: Frequency of Gales & Storms	Sea Surface: Monthly average Temperature °C Low	Sea Surface: Monthly average Temperature °C High	Sea Surface: Average Surface Salinity ‰	Sea Surface: Prevailing Current Direction (from)	Remarks (Productivity and Shelves)
1	**INDIAN OCEAN MONSOON REGIME**								
1.1	ARABIAN SEA REGION (with Red Sea and Persian Gulf)								
1.1.1	*Red Sea (N and S parts)*	Wi–NW/SE Su–NW	2–3	–	c.20	>30	40	Tidal	Oxygen minimum layer in c.75 m depth. Moderate production.
	Northern part	Wi–NW; Su–NW			20	28		Wi–SE	
	Southern part	Wi–SE; Su–NW			25	32		Su–NW	
1.1.2	*Persian Gulf*	Wi–NE; Su–W	<1	–	<15	32.5	39	Tidal	NW part has moderate production. Area 240 000 km². Av. depth 35 m.
1.1.3	*Gulf of Aden*	Wi–NE; Su–SW	1–2; 4–5	Wi –; Su +++	24	28	36	Wi–NE Su–SW	Oxygen minimum layer close to surface (50–150 m). Upwelling on the W coast during summer. Productive area.
1.1.4	*Gulf of Oman*	Wi–NW; Su–SW	<1; 3–4	Wi –; Su +++	22	30	36.5	Wi–N; Su–W	Very narrow shelf.
1.1.5	*Central Arabian Sea*	Wi–NE; Su–SW	3–4; 4–6	Wi –; Su +++	24	<26	36	Wi–NE Su–SW	Productive in central part along the oceanic current divergence. No shelf.
1.1.6	*Laccadive Sea* (Indian W coastal waters)	Wi–N; Su–W	2; 2–3	–	24	>27	35.5	N	North part of shelf productive. Area 300 000 km². Average depth NE part 20 m.
1.2	BAY OF BENGAL REGION								
1.2.1	*Bay of Bengal*	Wi–NE; Su–SW	2; 4	–	24	28	33	Wi–SE Su–SW	Oxygen minimum layer in c.150 m depth. Slight local upwelling on W coast during summer. N part has medium productivity. Area 170 000 km². Average depth: N part 15 m; rest 50–100 m.
1.2.2	*Andaman Sea*	Wi–N; Su–SW	1–2; 3	Wi –; Su +	26	27	32	Wi–N Su–W	Area 380 000 km². Average depth 80 m. Rocky and coralline. Low productivity.

Table 3.1 Cont'd.

No.	Names of the Regions and Subregions	*Winds*			*Sea Surface*				*Remarks*
		Prevailing Direction (from)	Resultant Force (Beaufort)	Frequency of Gales & Storms	Monthly average Temperature °C Low	High	Average Surface Salinity ‰	Prevailing Current Direction (from)	(Productivity and Shelves)
1.3	INDIAN OCEAN NORTH EQUATORIAL CURRENT REGION								
1.3.1	*Somali Waters*	Wi–NE; Su–S	>3; 3–6	Wi –; Su +++	23	27	35.5	Wi–NE Su–SW	Upwelling during the summer. Productive waters during this season.
1.3.2	*Indian Ocean N Equatorial Current Waters*	Wi–NE; Su–W	2–3	–	27	28	35	Wi–E Su–W	Slight upwelling along divergence and mixing on offshore grounds. Offshore shelf area 240 000 km². Average depth 40–80 m.
1.4	INDIAN OCEAN EQUATORIAL COUNTER CURRENT REGION	Wi–NW; Su–SE	<1; 1–2	–	26	>28	34.5	W	
1.5	INDIAN OCEAN SOUTH-EQUATORIAL CURRENT REGION								
1.5.1	*Mozambique Strait*	Wi–E; Su–SE	1; 2	–	21	28	35	N	Thick, nutrient-poor surface layer. Unproductive area. Madagascar shelf 180 000 km²; 15 m deep. Zambesi shelf 60 000 km²; 30 m deep.
1.5.2	*Indian Ocean S Equatorial Current Waters*	Wi–E; Su–SE	3	++	24	28	35	E	Unproductive area.
1.5.3	*North Australian Waters* (included: Timor Sea and Arafura Sea)	Wi–W; Su–E	1–2; 2–3	–	26	>28	34.5	Wi–SW Su–E	Low productivity. Arafura shelf 1 350 000 km². Average depth 70 m. NW Australian shelf 300 000 km². Average depth 80 m.
1.5.4	*Agulhas Waters*	Wi–SE; Su–W	>1; 5	Su +	16	23	35.5	NE	Low productivity. Area 110 000 km². Average depth 50–100 m.
1.6	INDIAN OCEAN HORSE LATITUDE REGION								
1.6.1	*Indian Ocean Southern Cyrals*	Wi–SE; Su–quiet	1–2; quiet	Wi ++; Su +	16	22	>35.5	2 gyrals	Low productivity oceanic area. Slightly more production on the southern boundary (sub-tropical convergence).

1.6.2	*West Australian Waters*	Wi–S; Su–W	3; 3–4	Wi +; Su –	16	21	36	Wi–S; Su–N	
1.6.3	*Great Australian Bight*	Wi–Varying Su–W	3–4	Wi +; Su ++	12	18	>35	Wi–E; Su–W	Medium productivity in the eastern part. NW Australian shelf 350 000 km^2. Depth 40–130 m. Tasmanian Shelf 170 000 km^2. Depth 60–80 m.
2	**WESTWIND DRIFT REGIONS**								
2.1	INDO-PACIFIC WESTWIND DRIFT REGION	W	Wi– 4–5 Su– >6	+ +++	5	10	34.5	W	Area north of subantarctic convergence. Productive. New Zealand shelf 250 000 km^2. Depth 70 m.
2.2	ATLANTIC WESTWIND DRIFT REGION	W	Wi– 6 Su–4	Wi +; Su ++	5	10	34	W	
2.2.1	*Patagonian Waters*	W	Wi–<3 Su–4	++	7	15	33.5	Tidal (S)	Large, medium to high productive shelf area. Great seasonal variations. Area 1 000 000 km^2, depth 80–100 m.
2.3	ANTARCTIC NORTHERN REGION	NW	Wi–4–5 Su–6	Wi ++ Su +++	–0.5	2.5	34	SW	Never covered with pack ice. Important whaling area. Very productive.
2.3.1	*Area North of the Ross Sea*	NW	6	+++	–0.5	4	<34	(SW)	Southern part covered with pack ice during northern summer. Whaling area. Productive.
2.3.2	*Scotia Sea and South Georgian Area*	W	>6	Wi +; Su +++	0	2	34	SW	Never covered with pack ice. Important whaling area. Very productive.
2.4	ANTARCTIC INTERMEDIATE REGION	NE	4	+++	<–1	–1	<34	(SW)	} Largely covered by pack ice in northern summer and fall. Productive.
2.4.1	*Waddell Sea*	NW	4–6	+++	<–1	–1	>33.5	(Gyral(E))	}
2.5	ANTARCTIC SOUTHERN REGION	E	Wi– 3–4 Su– >6	Wi +; Su +++	<–1	<–1	<33.5	SE–E	Largely covered by pack ice throughout the year, with exception during the northern winter.

Table 3.1 Cont'd.

No.	Names of the Regions and Subregions	Winds: Prevailing Direction (from)	Winds: Resultant Force (Beaufort)	Winds: Frequency of Gales & Storms	Sea Surface: Monthly average Temperature °C Low	Sea Surface: Monthly average Temperature °C High	Sea Surface: Average Surface Salinity ‰	Sea Surface: Prevailing Current Direction (from)	Remarks (Productivity and Shelves)
3	**ARCTIC REGION**								
3.1	KARA SEA	(Wi–NE) (Su–SW)		Su–; Wi ++	<−1	4	<33		Largely covered with drift and pack ice, except during the summer. North Siberian Shelf 2 600 000 km^2. Depth: W part 25 m; E part 40 m.
3.2	NORTH SIBERIAN WATERS	(Wi–NE) (Su–SW)		Su +; Wi +++	<−1	1	325	(W)	Largely covered with pack ice. Low productivity.
3.3	CHUKTSCHEE SEA	(Su–SE)		Su –; Wi ++	<−1	1	<33.5	(NW–E part) (SW–W part)	Largely covered with pack ice except few months during late summer.
3.4	HIGH ARCTIC				<−1	<−1			Covered with ice year around.
4	**PACIFIC OCEAN**								
4.1	KAMCHATKA REGION								
4.1.1	*Okhotsk Sea*	Wi–N; Su–S	5; 1	Wi +; Su –	<−1	11	>32	(N)	Productive shallow area with great seasonal change. Area 580 000 km^2. Depth 140 m.
4.1.2	*Kamtchatka-Kurile Waters*	Wi–N; (NE–NW) Su–S	6 1	Wi +; Su +	0	12	32.5	NE	Southerly flow of cold, dilute, nutrient rich water. Mostly ice-covered in winter.
4.2	ALASKA REGION								
4.2.1	*Western Bering Gyral*	Wi–(centre of low) Su–S	>6 2	Wi + Su –	0	8	33	Gyral	Variable currents, slow counter-clockwise circulation. High nutrient content.
4.2.2	*Alaska Coastal Waters*	Wi–NE SU–SE	>6 2	Wi +; Su –	<−1	8	32	Tidal(SE)	Northerly flow of warm, dilute, medium nutrient-content water. Bering Shelf 1 200 000 km^2. Average depth in NE 35 m.
4.2.3	*Alaska Gyral*	(Wi–S); Su–S	6; 2–3	Wi +++; Su +	4	13	32.5	Gyral	Subarctic water that turns N-ward and forms a counter-clockwise gyral. Nutrient content high. Deep oceanic area.

4.2.4	*NW American Coastal Waters*	(Wi–S) Su–NW	<5; 1–2	Wi ++; Su –	13	15	<32.5	(SW) Su in S part NE	Northerly flow north of about 50°N and southerly in lower latitudes. Nutrients moderate to high.
4.3	NORTH CHINA AND JAPAN SEAS REGION								
4.3.1	*North China Sea*	Wi–N; Su–SE	4; 1	Wi –; Su +	10	27	33.5	Tidal (Wi–N; Su–SW)	Productive area. Great seasonal changes. Area 1 000 000 km^2. Depth: W part 30 m; E part 80 m.
4.3.2	*Sea of Japan*	Wi–NW; Su–S	4; 1	Wi +; Su +	5	18	34	SW	Moderately productive. Great seasonal changes.
4.4	NORTH PACIFIC DRIFT REGION	Wi–W; Su–SW	>6; 3–4	Wi +++; Su ++	7	15	<34	W	Productive oceanic area with westerly flow of relatively warm water. Deep oceanic area.
4.5	CENTRAL NORTH PACIFIC REGION								
4.5.1	*North Pacific Gyral Waters*	Wi–W; Su–(E)	Wi–4; Su–1	Wi ++; Su –	16	25	35	(W) (Gyrals)	Slow circulation. Unproductive. Deep oceanic area.
4.5.2	*S. Francisco Waters*	Wi–(NE); Su–N	Weak; 6	Wi +; Su –	15	18	33.5	NW	Upwelling along the coast. Productive area.
4.6	PACIFIC NORTH EQUATORIAL CURRENT REGION								
4.6.1	*Philippine Waters*	Wi–NE; Su–S/SE	>3; 1–2	Wi ++; Su +	20	>28	34.5	SE	Origin of Kuroshio. Low productivity.
4.6.2	*Pacific N Equatorial current Waters*	Wi–NE; Su–E	>3; 1–2	–	22	>28	34.5	E	Unproductive, except on divergences along counter current region. Deep oceanic area.
4.6.3	*California Coastal Waters*	Wi–N; Su–(N)	1; 2–3	–	16	23	34	NW	Moderate seasonal variations and moderate to high productivity.
4.6.4	*West Mexican Waters*	Wi–NE; Su–(NW)	1–3; <1	Su ++(+)	22	29	Wi >34.5 Su <33.5	SE	Small to moderate seasonal variations. Seasonal upwelling along the coast.
4.7	INDONESIAN REGION								
4.7.1	*South China Sea*	Wi–NE; Su–S	2–3	Wi +; Su +	20	>28	33	Wi–N; Su-S	Shallow sea, Gulf of Siam and NW coast relatively productive.
4.7.2	*Java and Flores Seas*	Wi–NW; Su–SE	1–2; 2	–	27	27	33	Wi–NW; Su–SE	South China shelf 2 300 000 km^2. Average depth 50 m.
4.7.3	*Sulu and Celebes Seas*	Wi–N; Su–SE	1–2	–	27	27	34	Wi–N; Su–SE	Unproductive area.
4.8	PACIFIC EQUATORIAL COUNTER CURRENT REGION	Wi–NE/E Su–SE	1–2	–	24(E part) 27	28 28	<34.5	W	Unproductive area except on divergences along counter current region. Oxygen minimum layer deep (500–1000 m). Currents vary seasonally.

Table 3.1 Cont'd.

No.	Names of the Regions and Subregions	*Winds*			*Sea Surface*				*Remarks*
		Prevailing Direction (from)	Resultant Force (Beaufort)	Frequency of Gales & Storms	Monthly average Temperature °C Low	High	Average Surface Salinity ‰	Prevailing Current Direction (from)	(Productivity and Shelves)
4.9	SOUTH PACIFIC								
4.9.1	*North Polynesian Waters*	Wi–NW/N Su–SE	1; 2–3	–	27	28	34.5	E	
4.9.2	*Coral Sea (NE Australian Waters*	Wi–E; Su–SE	2–3; >3	+	19	27	35	Wi–E; Su–SE	Unproductive. Queensland Shelf 220 000 km^2. Depth <40 m.
4.9.3	*N. Tasman Sea Waters*	Wi–E; Su–SE	3; 2–3	Wi +; Su ++	16	22	>35.5	(Wi–E) (Su–NE and SW)	Moderately productive.
4.9.4	*Peru-Galapagos Waters*	Wi–S; Su–SSE	2–3; 5–6	–	18	22	>35	SE	Very productive waters.
4.10	PACIFIC SOUTHERN GYRALS REGION								
4.10.1	*Pacific Southern Gyrals*	Wi–E; Su–SE	1; 2–3	Wi +; Su –	17	23	35.5	Gyrals (NE)	Low productivity. Deep oceanic area.
4.10.2	*SW Chilean Waters*	Wi–SE; Su–W	3; 6	Wi +; Su ++	10	17	34	SW	Moderately productive. Very stormy.
5	**ATLANTIC OCEAN**								
5.1	ATLANTIC SUBARCTIC REGIONS								
5.1.1	*E Greenland Waters*	NE	Wi–6; Su-3	Wi ++; Su –	–1	0	32	NW	Cold E Greenland current. Ice covered most of the year. Organic production high on southern boundary.
5.1.2	*Barents Sea*	Wi–SW; Su–N	6; 2	Wi ++; Su –	1	5	34	(SW)	Influenced by warm North Cape Current. Mixing zone productive. Covered with ice during winter. 850 000 km^2. Average depth: 200–300 m.
5.1.3	*Labrador Waters*	Wi–NW; Su–SW	4; 2	Wi ++; Su –	–1	6	34.5	N	Cold arctic water. Mixing zone productive. 160 000 km^2. 120 m depth.
5.1.4	*Baffin Bay*	Wi–N; Su–N	3; 2	Wi ++; Su –	–1	2	32	(N)	Cold area, covered with ice during greater part of year.

5.1.5	*Hudson Bay*				−1	3	28		Ice covered during greater part of the year. 1010000 km^2. Depth 60–150 m.
5.2	N ATLANTIC INTER-MEDIATE BOREAL REGION								
5.2.1	*New-foundland Waters*	Wi–W; Su–SW	4; 2	Wi +++; Su +	0	13	32	NW	Very productive area under the influence of cold Labrador current. Mixing area. New-foundland and Nova Scotia shelves 770000 km^2. Depth 60–150 m.
5.2.2	*Irminger Gyral*	SW	Wi–5; Su–2	Wi +; Su −	3	9	35	(NE)	Warm Irminger current. Productive in boundary regions (mixing).
5.2.3	*Norwegian Sea-Faeroes Waters*	SSW	Wi–6; Su–2	Wi ++; Su −	5	12	35	SW	North Atlantic Drift-Norwegian Current. Productive. Norwegian Shelf 93000 km^2. Depth 200–300 m. Iceland–Faeroes Shelf: 115000 km^2. Depth 200–300 m.
5.2.4	*N. Sea, Irish Sea, English Channel*	SW	Wi–3; Su–2	Wi ++; Su −	7	16	34	Tidal (SW) (NW)	Shallow area, medium productivity. North Sea 575000 km^2. Depth 94 m. English Channel 75000 km^2. Depth 54 m.
5.2.5	*Baltic Sea*	SW	2–3	Wi +; Su −	0	17	2–28		Baltic Sea 390000 km^2. Depth 100 m. Brackish water. Medium to low productivity
5.3	GULF STREAM-ATLANTIC DRIFT CURRENT REGION								
5.3.1	*Florida Waters*	Wi–W; Su–SE	1; 1	Wi +; Su −	23	28	36	S	Very strong current. Low productivity.
5.3.2	*Gulf Stream Waters*	Wi–W; Su–S	2; 1	Wi ++; Su −	15	25	35.5	SSW	Strong currents. Low productivity. New England-Carolina shelf 220000 km^2. Depth 35–80 m.
5.3.3	*Atlantic Drift Current Waters*	SW	Wi–2–3 Su–1–2	Wi +++; Su +	10	15	35.5	SW	Medium to low productivity except in N boundary. Deep oceanic area, except Gulf of Biscay. 80000 km^2. Depth 130 m.
5.4	CENTRAL NORTH ATLANTIC REGION								
5.4.1	*Sargasso Sea*	Wi; Su–E	(1); –	Wi +; Su −	20	27	36.5	Gyral	Unproductive. Deep oceanic area.
5.4.2	*Azoren Waters*	NNE	1(F); 2(A)	Wi +; Su −	17	22	36.5	NNE	Unproductive area (Bay of Cadiz is slightly more productive). West Pyrenean 50000 km^2. Depth 250 m.

Table 3.1 Cont'd.

No.	Names of the Regions and Subregions	*Winds* Prevailing Direction (from)	Resultant Force (Beaufort)	Frequency of Gales & Storms	*Sea Surface* Monthly average Temperature °C Low	High	Average Surface Salinity ‰	Prevailing Current Direction (from)	*Remarks* (Productivity and Shelves)
5.5	MEDITERRANEAN								
5.5.1	*Mediterranean* Western Basin	Wi–NW; Su–N	Weak 1; 1	Wi +; Su –	13	24	37.5	S part–W N part–E	Strait of Gibralter and western part productive, rest unproductive. Balearic and Gulf of Lion 40 000 km^2. Depth 150–100 m.
	Eastern Basin	Wi–W; Su–NW	1; 2	Wi +; Su	15	28	38.5	No tides	Adriatic Sea is shallow, cooler and less saline. Aegean sea also cooler. Adriatic Shelf 90 000 km^2. Depth 30–150 m. Sicilian-Sidra shelf 130 000 km^2. Depth 50–100 m.
5.5.2	*Black Sea*				5	23	14–28	No tide	Brackish surface, H_2S in deep water. Azov sea was very productive.
5.6	ATLANTIC NORTH EQUATORIAL CURRENT REGION								
5.6.1	*Gulf of Mexico*	Wi–NE; Su–W	Wi–2; 1	–	22	29	>36	N part-gyral SE part-SW	Low productivity. Small seasonal variations. Florida-Texas shelf: 450 000 km^2. Depth 40 m. Shallow coast. Campeche 180 000 km^2. Depth 30 m.
5.6.2	*Bahamas Waters*	E	Wi–1–2 Su–2	–	24	29	36.5	NW	Shallow area with great banks. Small seasonal variations. Very low production. Bahamas shelf 130 000 km^2. Depth 15 m.
5.6.3	*Caribbean Waters*	E(Trades)	2–4	–	26	28	<36	E	Low productivity, except SE part, where minor upwelling occurs. Mosquito 110 000 km^2. Depth 25 m. Venezuela 130 000 km^2. Depth 20–80 m.

5.6.4	*Atlantic N Equatorial Current Waters*	NE	Wi>5; Su>3	–	26	27	35.5	E	Unproductive oceanic area. Amazon 540 000 km². Depth 15–70 m.
5.6.5	*Cape Verde Waters*	NNE	Wi–3; Su–3	–	21	24	36.5	NE	Slight upwelling in several locations along the coast. Productive along the coast and on SE boundary.
5.7	GUINEA REGION	Wi–S and N Su–SSE	2	–	24	27	34.5	NW; W	NE boundary on the Sierra Leone shelf is productive. The rest of the area low productivity. O_2 minimum layer in c.100 m depth. Sierra Leone, Guinean shelf 210 000 km². Depth 25 m.
5.8	ATLANTIC S EQUATORIAL CURRENT REGION								
5.8.1	*Atlantic S Equatorial Current Waters*	SE	>3	–	24	27	36.0	E	Low productivity oceanic area. O_2 minimum layer close to surface (c.50 m).
5.8.2	*E Brazilian Waters*	SE	Wi–3; Su>4	–	24	27	>37	NE	Low productivity except around Cabo Frio. S Brazilian shelf 400 000 km². Depth 80 m. (5.8.2–5.8.3)
5.8.3	*SE Brazilian Waters*	NE	1–2	–	19	24	36.5	N	
5.8.4	*Benguela Current Waters*	SE	Wi>3; Su>4	–	17	22	36.0	SE	Oceanic area with medium production.
5.8.5	*SW African Waters*	SSE	3	–	16	21	35.5	S(SSE)	High productivity area. Extensive upwelling along the coast. O_2 minimum layer close to surface. St. Helena shelf 140 000 km². Depth 150 m
5.9	ATLANTIC SOUTHERN GYRALS REGION	(Often quiet)	1–2	–	15	20	35.5		Low productivity oceanic area.

NOTES TO TABLE 3.1

Water temperatures: (1) The first number refers to the average temperature of the coldest month and the second to the average of the warmest month; (2) Water temperatures are rounded off to the nearest full °C. *Winds and currents*: (3) Abbreviations for seasons; Wi-Northern winter (average for month of February); Su-Northern summer (average for month of August); At-Northern autumn; Sp-Northern spring. (4) Resultant force of wind and average speed of currents: first value refers to northern winter, second to northern summer. (5) Frequency of gales and storms: – nearly absent or rare; + moderately frequent; ++ frequent; +++ very frequent. (6) Use of parenthesis. The values in parenthesis give no representative average of the area or a second choice of the season or value appropriate for the greater part of the area.

Table 3.2 Groups of natural regions of the oceans with similar environmental conditions

		Groups of similar regions (for names see Table 3.1 and for locations see Fig. 3.1)	
Polar	High Polar regions	Group I	2.5; 3.4
		Group II	3.1; 3.2; 3.3
Boreal	Subpolar regions	Group III	2.4; 4.1; 5.1
		Group IV	4.2; 5.2; 2.3
Boreal	Westwind Drift regions	Group V	2.1; 2.2; 4.4; 5.3.2
	Subtropical Gyral	Group VI	4.5.1; 5.4.1; 4.10.1; 5.9.1; 1.6.1
	Eastern upwelling regions	Group VII	4.5.2; 5.4.2; 5.8.5; 4.9.4
		Group VIII	4.6.3; 4.6.4; 5.6.5; (1.5.3?)
		Group IX	4.10.2; (1.6.2)
Tropical	Equatorial Current regions	Group X	4.6.2; 4.9.1; 5.8.1; 1.5.2
	Western Subtropical Gradient Current regions	Group XI	5.6.1; 5.6.2; 5.6.3; 5.8.2; 1.5.1; 1.5.6; 1.3.1; 4.9.2; 4.9.3; 4.6.1
	Equatorial Counter Current regions	Group XII	1.4; 4.8; 5.7
	Indo-Pacific Monsoon regions	Group XIII	1.1; 1.2; 1.3; 4.7
		Group XIV	4.3; (5.3.1)

The boundaries of these oceanic regions must often be considered as relatively wide transition areas. They are often boundaries of different currents or different water masses and/or surface water types. In some cases, the boundaries are drawn in regions with high horizontal temperature gradients and therefore the actual boundaries would shift slightly with the season. As the regions refer only to surface water types, no consideration has been given to the structure of the sub-surface waters or to the determination of boundaries by the submarine features.

Each hydroclime region can be characterized by its environmental conditions (properties). One of the most used properties is water temperature, on which most data and longer time series observations are available. Temperature, however, might not be ecologically the most important property of the environment. Surface currents are significant in several aspects of ecosystem dynamics. Thus many natural regions in the oceans are characterized and delineated by current systems.

Changes in salinity are small and usually indicative of advective changes and mixing. Exceptions to this are coastal areas and estuaries where run-off affects salinity. Of other chemical properties, the changes of nutrient salts (mainly phosphates and nitrates) are indicative of productivity changes and eutrophication. Likewise, changes in light conditions (cloudiness, turbidity in the water) might affect basic organic production as well as fish behaviour. Most changes of biological conditions and processes are subject to the study of the response of the ecosystem to changes and anomalies of the hydroclime.

Of special concern with processes are the magnitudes of rates of change and the timing of seasonal changes in the ocean. These processes usually reflect the speed of response of the hydroclime to marine climatic changes. The variation in timing of seasonal change is especially important in evaluating the environmental effects, as the difference in timing and/or in rate of seasonal change will appear as an anomaly of a given property (e.g. temperature).

In evaluating the effects of climatic and hydroclime changes on marine fish ecosystems, some prior knowledge of whether and how the environment affects and determines the distribution and abundance of marine resources must be at hand. In the past, correlations between a single environmental parameter (e.g. temperature) and a biological parameter (e.g. catches of a given species) have been abundantly used. However, these correlations do not explain the mechanisms of interactions and usually fail to prove that climatic change has been the cause for change within fish ecosystem, and so many of the selected correlations have been spurious.

Many attempts have been made at direct empirical correlation between recruitment (and/or year class strength) and environmental factors, most of which have failed, as recruitment is a complex process and might be largely determined within ecosystem, e.g. by predation. In many cases of empirical correlation only a conjecture about the relationship between fish and environment can be put forward and little has been established with certainty. We must know why and how an environmental factor affects fish abundance and attempt to quantify this effect. We must also recognize normal biological changes within the ecosystem such as competitive species interactions, which can at times be complex. Thus only a cursory attempt can be made in this book to review critically the environment–fish interactions from the point of view of the climatic change effect.

It has been shown that a dominant species in an ecosystem responds differently from the subdominant species to the environmental changes (Skud 1982). The response of predator species can also be different from that of predominantly prey species. This has been demonstrated by Laevastu (1984) with pollock and capelin in the Bering Sea. Figure 3.2 shows the change of pollock biomass in Bristol Bay caused by two different temperature anomalies in three consecutive years. Temperature in this case affects the growth rate and also food requirements of pollock. Corresponding changes of capelin and other small pelagic forage fish are shown in Fig. 3.3. The changes in the pelagic forage species biomasses are opposite to the changes in pollock biomass, and are caused by predation by pollock, which is the predominant species in this ecosystem. It should also be noticed that the total temperature anomaly effect with very large anomalies in three consecutive years (1.5, 2.5, 1.5 °C) is less than 20% of the original biomass.

The consumption of prey species by predators, which in most cases are

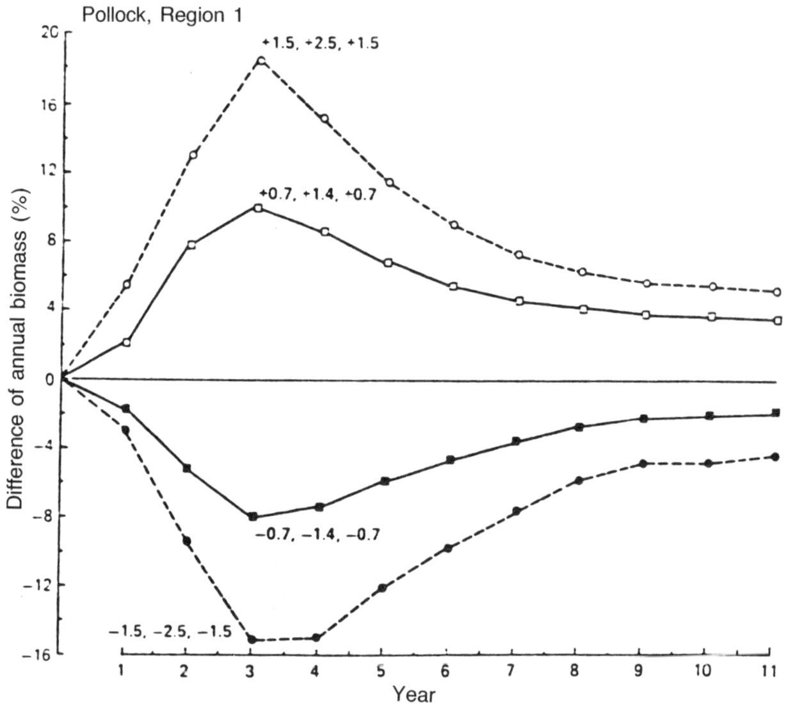

Fig. 3.2 Annual differences of biomass of walleye pollock in Bristol Bay, Bering Sea, caused by 3-year anomalies of temperature.

important commercial species such as cod, can be considerable and variable from year to year. Mehl (1989) found that cod in the Barents Sea were in 1985 eating about 2 000 000 t of capelin and in 1986 less than half of this amount. Considerable quantities of redfish, herring and haddock were also consumed. Magnusson & Pálsson (1989) found that predation and fishing of capelin in Icelandic waters were of a comparable order, and that the catch and predation estimates showed a discrepancy with acoustic stock-size estimates of the order of two.

3.1 Seasonal changes and anomalies in relation to climatic changes

Seasonal changes occur in environmental properties in medium and high latitudes of the oceans. In this section the characteristics, including magnitudes, of these seasonal changes are briefly compared with possible climatic changes. Figure 2.15 shows the magnitude of annual surface temperature change in the North Atlantic (Schott 1944). As seen from this figure the seasonal changes in many regions are higher than any known corresponding

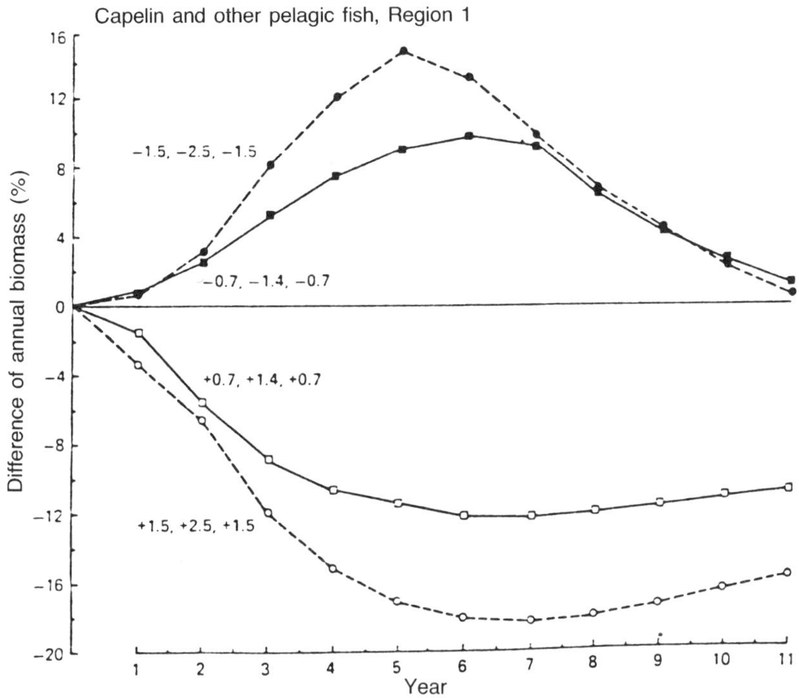

Fig. 3.3 Annual differences of biomass of capelin and other pelagic fish in Bristol Bay, Bering Sea, caused by 3-year anomalies of temperature.

long-term climatic change. It should be noted that large monthly and seasonal temperature anomalies can arise in areas of large seasonal changes if the timing of the annual change is unusual.

Monthly means and anomalies, of e.g. temperature, in a given location can vary considerably from one year to another as well as from one decade to another. An annual anomaly can be caused by anomalies in few months only, i.e. by seasonal anomalies. The wind and surface temperature anomalies are most often functionally related. Rodewald (1971) found that in 1961–70 cooling of the North Atlantic was mainly a summer phenomenon of the order of −0.5 °C (Fig. 3.4). The cooling tongue extended from Nova Scotia in the general direction of the Polar Front to about 60 °N, 10 °W. However, the Irminger Sea showed a slight warming trend in the winter, whereby isallotherms resembled the isobars of the normal Icelandic Low. Rodewald (1971) furthermore found that the temperature changes during 1961–70 at coastal stations (St Andrews, N.B., and Boothbay Harbor, Maine) were quite different and often opposite to the changes at the nearest Ocean Weather Station (OWS) *D*. The cooling trends in coastal stations were

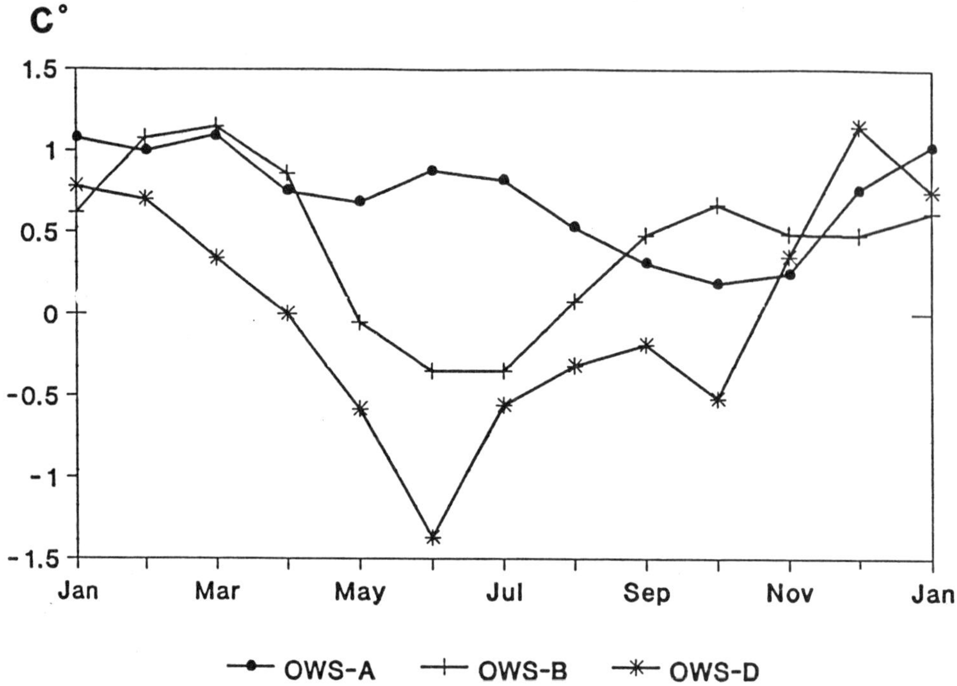

Fig. 3.4 Deviation of 1961–70 monthly means of sea temperature for ocean weather stations *A*, *B* and *D* (Rodewald 1971).

related to changes of the offshore wind component, and the offshore sea temperature changes were also largely dependent on the type and strength of atmospheric circulation.

Time series of oceanographic measurements are the bases for evaluation of climatic and hydroclime fluctuations. The best time series are available from the North Atlantic, most of which have been evaluated by Jens Smed in his numerous reports to ICES (e.g. Smed 1980, 1983, Smed *et al.* 1983). These evaluations show relatively minor ($< \pm 0.5\,°C$) variations of sea surface temperature in the North Atlantic, variable from one region to another but with a general positive anomaly in the late 1950s and early 1960s, and deterioration in the late 1960s and early 1970s. Dickson & Lamb (1972) pointed out an increased southward transport of pack-ice (Fig. 3.5) and active ice formation off North Iceland in the late 1960s. On the other hand, the sea surface temperature anomalies seem to be somewhat larger in the North Pacific (Fig. 3.6), change more rapidly over a few years and vary with latitude (Favorite & Ingraham 1976).

Fish ecosystems have many season-dependent processes such as spawning and feeding migrations, migration from deep to shallow water, hatching of

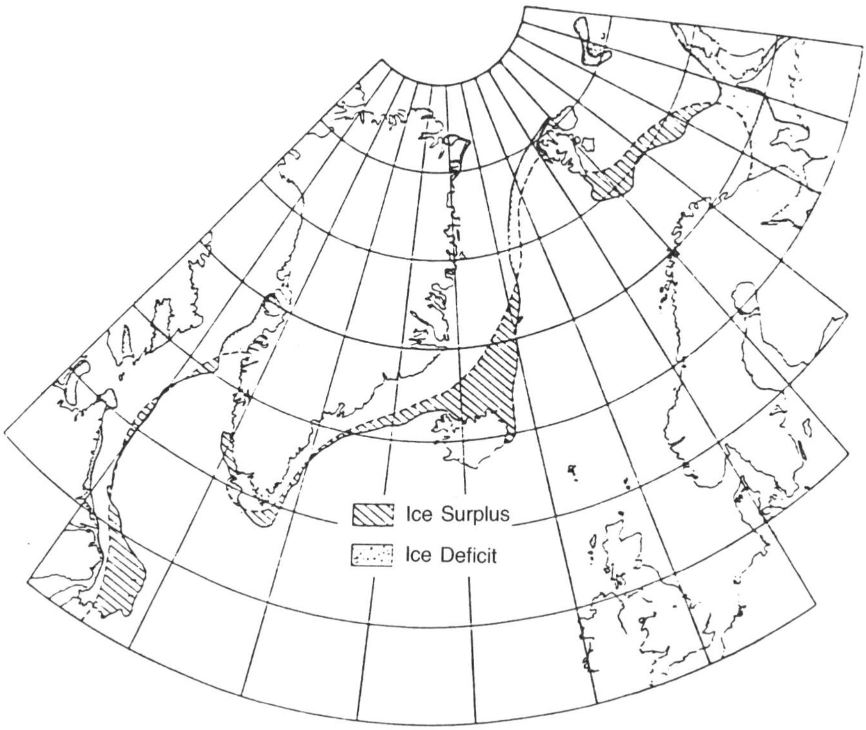

Fig. 3.5 The extent of ice at 8 May 1968 (solid line) compared with the normal (broken line). The normal is a composite based on an American average concentration of ice 1911–50 (Dickson & Lamb 1972).

larvae, etc. It is often difficult to separate innate seasonal behaviour and the behaviour triggered and/or modified by environmental changes. The timing of seasonal changes of environment often gives a clue to the reaction of fish to the environment. Favorite & Laevastu (1981) described the seasonal changes of environment in the eastern Bering Sea and the accompanying seasonal movement of fish, both active migration and passive drift of larvae.

The behaviour of fish undergoes seasonal changes of various kinds, including seasonal migrations. Thus the availability of fish for capture and the species composition of catches will vary seasonally (Fig. 3.7) (Klimaj 1976). Some of the seasonal behaviour (e.g. seasonal migration) is affected by the seasonal changes of environment, but much of the seasonal behaviour is innate. Examples of innate seasonal depth migrations of two species are shown in Fig. 3.8. Alverson (1960) concluded that seasonal vertical migrations result in seemingly paradoxical seasonal change in the productivity of the species complex. The highest individual catches for many species are made during the winter months, but the highest aggregate catches for the species complex

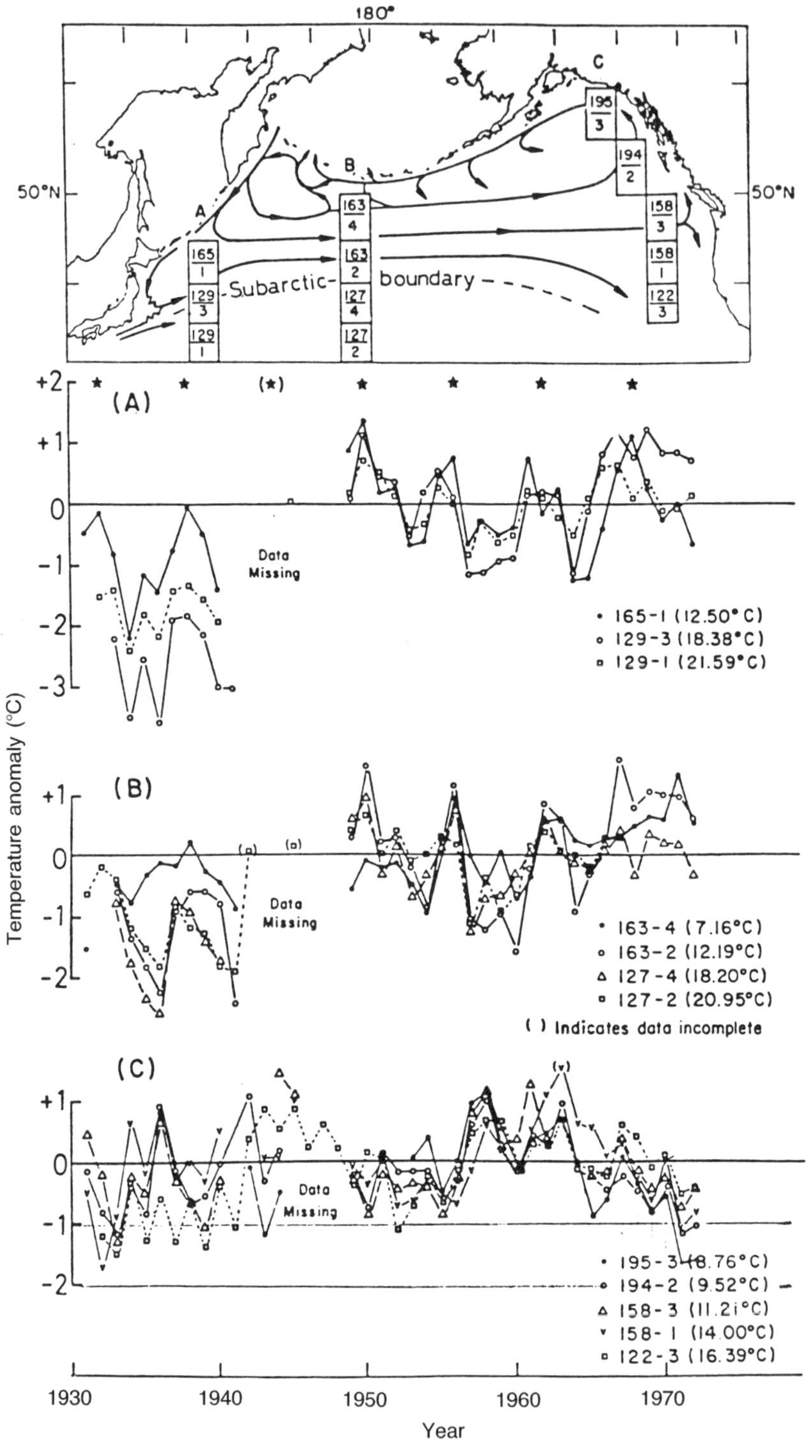

Fig. 3.6 Deviations from the annual mean sea surface temperatures in the numbered 5 × 5 degree quadrangles, 1930–72 indicating the long-term variability and the transpacific continuities and discontinuities of specific events (Favorite & Ingraham 1976).

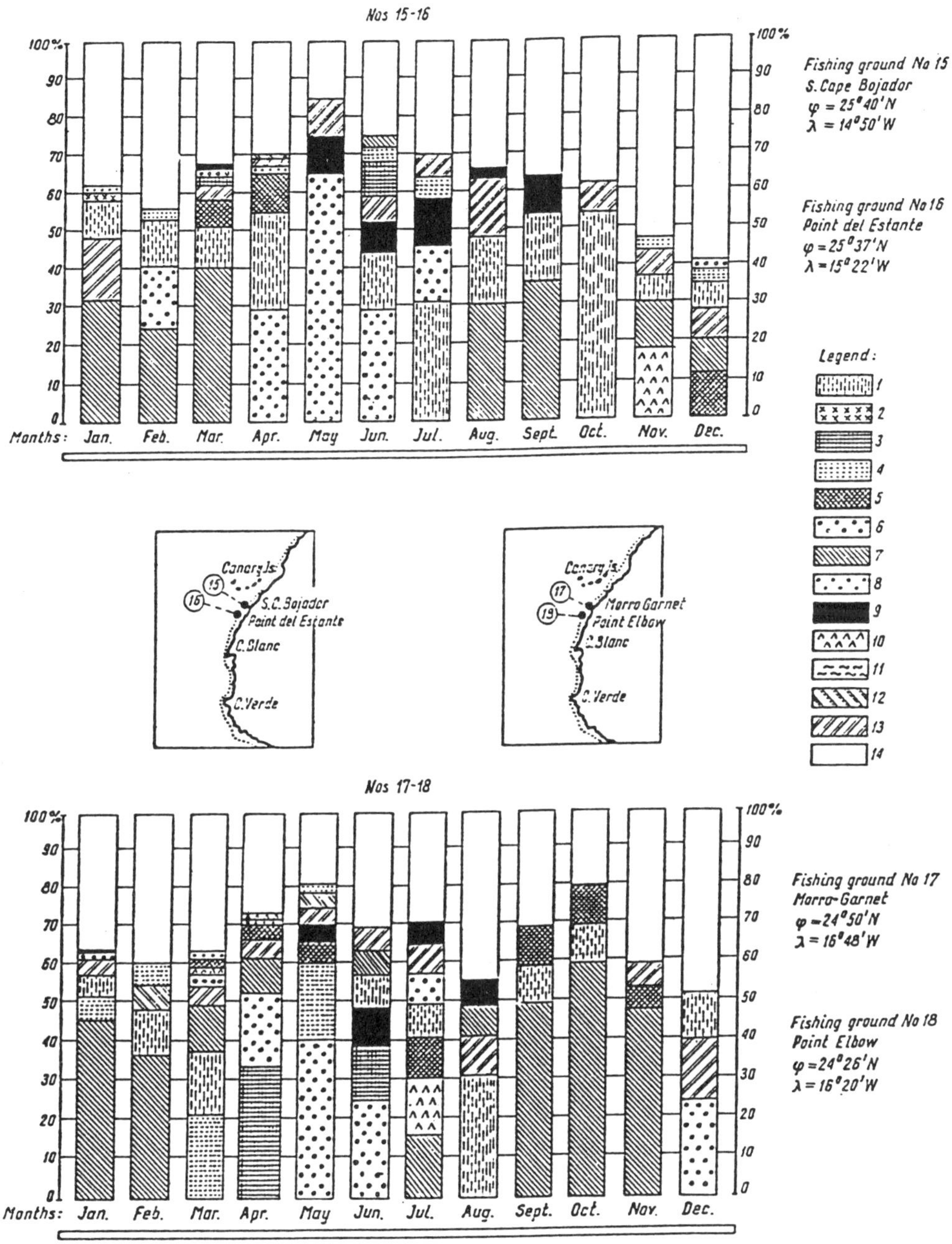

Fig. 3.7 Mean daily bottom trawl species composition (%) for catches taken from the Northwest African shelf grounds by the B–18 and B–23 freezer trawlers in the years 1965–9 (annual cycle): 1-horse mackerel; 2-*Decapterus rhonchus*; 3-'lichia'; 4-dogsteeth; 5-hairtail; 6-alache; 7-chum mackerel; 8-hake; 9-bluefish; 10-Atlantic bonito; 11-gilthead bream; 12-barracuda; 13-other fish; 14-fish meal (Klimaj 1976).

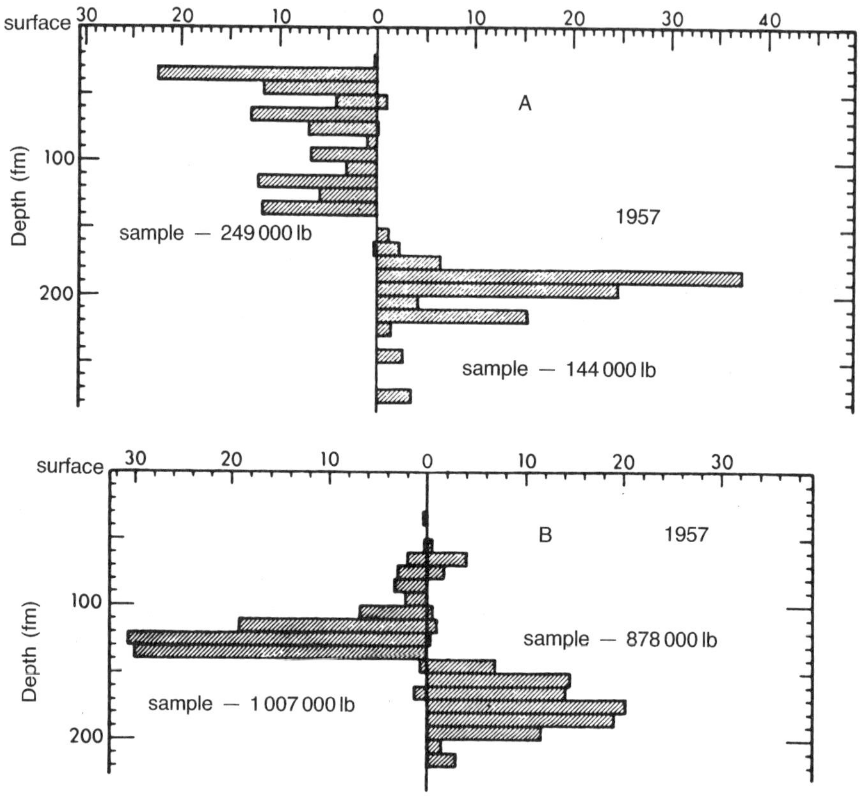

Fig. 3.8 Percentage of total summer and winter catches of A Dover sole and B Pacific ocean perch in 10 fm depth intervals off the Washington and Oregon coasts in 1957 (Alverson 1960).

are taken during the summer. Thus knowledge of the bathymetric movements should play an important role in the understanding of the dynamics and management of fish ecosystems.

The seasonal behaviour also changes with age of the fish. For example, juvenile cod in Icelandic waters are more numerous in shallower water during summer (June to September) than during winter. This seasonal pattern decreases with increasing age, and age-group 4 is found in greatest numbers at 200 to 300 m depth in all seasons (Pálsson 1984). Whether and to what extent seasonal anomalies of environment affect the seasonal behaviour is little known for most species. The seasonal occurrence of some species (e.g. squids in NE Japan) is related to seasonally prevailing water type and the boundary between warm and cold water currents. Thus their abundance and dominance in given location can vary. For example, common squid (*Todarodes pacificus*) was predominant during the summer seasons of 1968 and 1969, whereas *Ommastrephes bastrami* dominated in 1973 and 1974 (Murata *et al.* 1976). However, no clear relationship to water type and/or temperature could be demonstrated, as the general abundance and recruitment could also have varied between years.

Sjöblom (1978) found that spring-spawning Baltic herring in Finnish waters are most abundant following severe winters (1926–30, 1940–50) and autumn spawners during milder winters (1931–9, 1951–7). He ascribed the reason not directly to environment but to the abundance of plankton, which is more abundant during summers following severe winters.

In many interdisciplinary studies it is common to attempt to find correlation between phenomena and resulting conditions in the different disciplines. Correlation studies are especially numerous in fisheries oceanography, for example, possible relationships between the surface temperature at a coastal station and landings of fish along the coast. Most fisheries biologists assume that physical climate is important for ocean ecology. However, few quantitative and persisting relationships are found by considering single physical parameters (e.g. temperature) and a population parameter (e.g. annual catch). Seldom, if ever, has thought been given to possible mechanisms of these correlations. Indeed the field of fisheries oceanography has been simplified and banalized with these often meaningless correlation studies.

Radovich (1982) analysed what has been learned from the collapse of the California sardine fishery and showed that many simplified generalizations were presented, e.g., that climatic (temperature) changes were responsible for the replacement of the sardine by the anchovy. However, considerable north–south migrations and shifts of populations occurred which had no relationship to temperature. In fact, there were population shifts between different years quite opposite to those which might have been deduced from temperature anomalies. There were variable but independent spawning successes in different areas. Furthermore, there were considerable changes in the predator populations.

Although various correlation studies in fisheries oceanography will undoubtedly continue, it seems to be necessary to emphasize the importance of selecting variables in these studies which would be meaningfully related through cause and effect. Surface temperature and demersal fish are rarely related, unless surface temperature can be considered as an index for bottom temperature or for changes in currents.

The scale of ocean variability is another factor requiring serious consideration. Little value can be given to monthly and/or seasonal anomalies if short-term (days or weeks) fluctuations have considerably greater magnitude than those monthly anomalies.

Progress in fisheries oceanography studies can be made when they are based on known cause–effect principles and create continuity in space and time. This can be done with numerical simulation studies which consider the total environment and total biocoenosis together — i.e., total (or holistic) ecosystem simulations.

More information about environment–fish interactions can be obtained if two- or three-dimensional space and time distributions of environmental and biological ecosystem components are taken into consideration, as partly demonstrated by Beverton & Lee (1965), who considered fish food distribution as well as environmental properties. These researchers found that cod in the Spitsbergen area were confined in the relatively warm Atlantic water above 1.5 to 2.0 °C in the boundary water. Furthermore, they found evidence that the occurrence of local concentration of a preferred food (e.g. herring) within an otherwise tolerable and featureless hydrographic area caused aggregation of cod in them.

Seasonal changes of quantities landed and species composition of landings are normal in fisheries. The main reason for these seasonal fluctuations of landings are seasonal availability of given species to fisheries, for example during spawning or during river runs (salmons). However, large year-to-year differences in seasonality of landings occur, which might be related to resource availability and/or abundance, but might also be caused by changing fishing conditions (e.g. frequency of storms), market needs and switching target species to improve earnings.

3.2 Changes of surface pressure, wind systems and storm tracks

The major climatological surface pressure and wind systems (e.g. the Iceland and Aleutian Lows) change their intensity seasonally and their mean positions move north or south. These seasonal changes also have annual and longer term anomalies. More prolonged northward displacements of wind and pressure systems (belts) are usually associated with intensifications of surface weather (e.g. winds) and are also related to hydrographic events of which they are driving forces (Dickson & Lamb 1972), such as surface currents and advective temperature changes. These pressure and circulation anomalies can be presented in various ways. Selecting different normal or base periods results in different deviations of pressure fields.

The climatological mean surface pressure systems are created by summation of synoptic surface processes which depict the movement of the surface cyclones (lows). These storm tracks can vary considerably from year to year in space and time (Figs 2.2, and 2.3). The consequences of these variations are manifold; first they cause changes in the surface layers in the oceans, especially currents and mixing, which, in turn, will affect some components of the marine ecosystem. Secondly, they affect the fishing operations. For example, in March 1965 the waters over Grand Banks were considerably stormier than in 1967, whereas March 1967 was stormy in the Barents Sea.

The surface winds largely determine the mean position of the oceanic Polar Front, which in most areas coincides approximately with the mean atmospheric Polar Front. The year-to-year and season-to-season variation of the position of Polar Front can exceed 200 nautical miles. It has been postulated that the waters on both sides of Polar Front affect the feedback from the ocean to the atmosphere and steer the surface atmospheric systems to some extent. Rossov & Kislyakov (1969) hypothesized that if the position of the polar front is abnormally southern in summer, the storms in the following winter will have abnormally southern trajectories, whereby winter temperatures over most of Europe will have negative anomalies. An abnormally northern position of the polar front results in shifting of cyclones along more northern trajectories and in positive temperature anomalies over Central and Northern Europe.

Monthly surface wind anomalies can be computed using daily surface wind from the surface pressure distribution and forming long-term monthly means (e.g. from 1946 to 1985) from which the given monthly mean wind (also computed from daily analysis) can be subtracted. The resulting anomaly patterns are markedly cyclonic or anticyclonic (see example, Fig. 1.2) and relatively large in scale, although they can also change rapidly and profoundly from one month to another.

The surface wind anomalies can have various effects on the ocean, besides causing anomalies in surface currents. Cushing (1988) described the formation of the Great Salinity Anomaly of the 1970s as being caused by the stress of northerly winds off East Greenland in winter during the 1960s and drifting across the North Atlantic for nearly twenty years. He believed that this anomaly caused the reduction of primary and secondary production and reduced the recruitment of several commercial species.

The upwellings near the coasts are created by prevailing wind systems, and thus are sensitive to local wind anomalies. The El Niño off the Peruvian coast can also be considered as being caused by surface wind anomaly, which reduces the normally intense wind-induced upwelling off this coast. The 1982–3 El Niño caused a positive temperature anomaly in excess of 5 °C, limited the primary production and caused large changes in the fish biomasses (Arntz 1986).

3.3 Ocean surface currents, their changes and effects on fish distribution, migrations and recruitment

One of the main effects of surface wind and its anomalies on the ocean is the creation of surface currents. The wind-driven component of the surface current predominates in most areas except in the region of permanent density-driven currents, which are also modified by wind, and tidal currents in coastal areas, where, however, the net transport is also affected by wind.

Surface currents affect the advection of different water types, which can change the characteristics of the environment in a given location. Ocean surface water types are often characterized by temperature and salinity.

Both are non-conservative properties at the surface and subjected to local changes. Water colour and especially content of plankton have also been found a suitable surface water type indicator (Marumo 1957) and might be useful from the point of view of the ecosystem.

Currents will affect the migrations of fish by passive transport of juveniles from spawning grounds to nursery grounds, and might serve as a means of orientation of counter-current migration of adults from feeding grounds to spawning grounds. Thus surface current anomalies might affect both the distribution of larvae and juveniles and spawning migrations of adults. The distributions of major fish stocks are usually confined to given current systems, which are considered as gyrals. Many distributions and migrations of fish do not, however, fit into the scheme of gyral distribution and active counter-current migrations, such as that of the Icelandic capelin (Fig. 3.9).

Surface current anomalies affect the positions of frontal zones of surface temperature. These frontal zones have been found to affect fish distribution, which is often assumed to be related to temperature but can as well be related to current and/or water type. For example, Sveinbjörnsson *et al.*

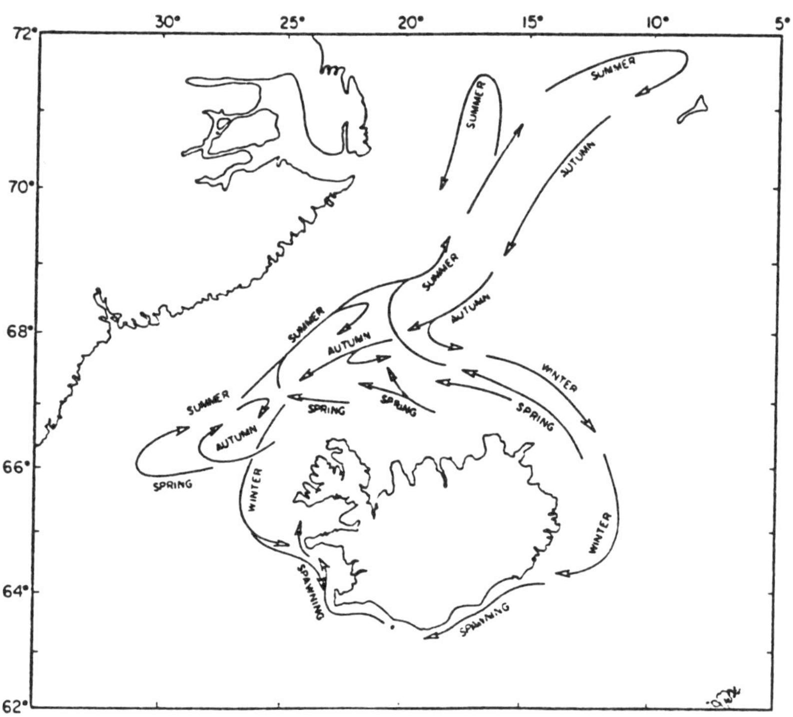

Fig. 3.9 The general migration pattern of the maturing capelin stock from April/May until March/April in the following year (Vilhjalmsson 1987).

(1984) found that in the area between Iceland and the Faeroes blue whiting were encountered on the warm water side of the frontal zone and very little migration had taken place in the cold water north of the Iceland–Faeroe Ridge in June. Other short-term movement related to surface current was noticed by Beverton & Lee (1965) pertaining to the movement of cod onto the western edge of the Barents Sea shelf as related to the surge of Atlantic water.

The association of pelagic fish such as tuna with current transition zones has been described by Laurs *et al.* (1977). They found that when the East Pacific transition zone (Sub-Arctic/Sub-Tropical) is distinct, albacore migrate along a corridor and are concentrated, often remaining in the transition zone for some time. When the boundary is diffuse, migration is over a broader area, and albacore move rapidly towards the US west coast. Reasons for this behaviour of tunas are, according to researchers, temperature preference, aggregation of food and thermal gradients, singly or together. Prediction of the position and structure of the transition zone would thus permit operational fishery decisions in the tuna industry.

The association of a stock of fish with a water type (mass) has been described by Seckel & Waldron (1960).

> 'The Hawaiian Islands are located in the vicinity of a boundary separating two water types. To the north and west is the western North Pacific type, with a salinity of more than 35‰. To the south and east is the transition type of the California Current Extension, with a salinity of approximately 34.7‰. It was found that this boundary, which is located just south of the islands during the winter months, begins to move northward early in the year. It passes through the islands in April or May and brings the lower salinity water of the California Current Extension. This process is reversed in August, when the boundary begins its southward retreat. It then passes the islands in October or November and brings the high salinity water of the western North Pacific.
>
> The spring movement of the boundary, bringing California Current Extension water into the island area, appears to be associated with the success of the Hawaiian skipjack fishery. Whenever California Current Extension water bathes the islands during the summer, as indicated by salinities of less than 35‰, the annual skipjack landings have, with the exception of 1958, been average or better than average. Whenever the salinities were higher than 35‰, as in 1952 and 1957, the annual landings were less than average.
>
> From these results one can postulate that commercial quantities of skipjack appear in the island area together with the California Current Extension. One can also postulate that there are years in which the California Current system fails to intensify sufficiently to move the

boundary into and through the island area. During such years, fishing is likely to be poor.'

In another study Seckel (1972) showed that skipjack travel from the eastern North Pacific to Hawaii by drifting in the North Equatorial Current.

Different water types have different plankton contents, both by abundance and by species dominance. It is possible that pelagic fish are associated with different plankton as food in these water types or with different food abundance, and are advected with these preferred water types.

Large scale changes of types of water masses caused by circulation changes have been assumed to cause changes in fish distribution. Russell *et al.* (1971) described such changes in the English Channel off Plymouth. They found, among other changes, that from the 1940s to 1964 pilchard eggs were abundant. Then in 1964 to 1971 pilchard eggs disappeared and herring, cod and ling appeared. These authors suggested that warming of the Arctic would affect circulation in the North Sea and let Atlantic Ocean waters extend further north, causing the observed changes. However, no such general circulation change has been found and/or demonstrated.

Laevastu *et al.* (1986) analysed the surface currents in the North Pacific from 1946 to 1984, using daily surface wind data. Although marked anomalies of surface layer transport occurred in individual months and years, it was not possible to demonstrate that these transport anomalies had caused fluctuations in stock abundance of the major commercial species in the North Pacific, including Bering Sea, because the fisheries data have large uncertainties and recruitment variations from year to year can also be caused by processes such as predation and their changes within ecosystems. The main difficulties in investigation of the environmental effects on recruitment are associated with the probable variability of a recruitment window. In fish such as walleye pollock, which spawn over a large area and over a long time period, the high probability of at least one recruitment window may result in small recruitment fluctuations from year to year.

Although the anomalies in conditions in the surface layers of the ocean can be constructed using wind as a driving and causative force, fisheries data from the North Pacific are inadequate indicators of changes in fish stocks, and consequently the effects of environmental anomalies cannot be assessed. However, some quasi-cyclic changes in surface layer temperature in the Bering Sea with periods of 7 to 11 years are apparent, and temperature fluctuations, similar but opposite in sign, have been found in the Barents Sea by Saetersdal & Loeng (1983). In both areas the major causes of temperature fluctuations have been found in surface wind anomalies. In the Barents Sea it has been shown that good year-classes of cod arise when the temperature cycle is increasing. This change usually means advection of Atlantic water into the Barents Sea, which might have two different effects.

Firstly, it can bring zooplankton eggs and nauplii with it and favour higher zooplankton production, and secondly, higher temperatures might favour better zooplankton as well as larvae growth. Phytoplankton production might also be enhanced by higher temperature and by mixing of nutrient-richer cold water with warmer Atlantic water. The possibility of enhanced transport of cod larvae and juveniles from spawning areas off Lofoten can also not be discounted.

The review in this subsection again underlines the uncertainties to concluding that regional climatic and hydroclime fluctuations might be considered as primary causes of catch fluctuations or stock abundance changes.

3.4 Temperature as an indicator of changes in the sea and a factor in fish ecosystem changes

Sea surface temperature is one of the ocean environmental parameters which is routinely observed and therefore numerous analyses, one to three dimensional, of it are available, including computations of anomalies (Fig. 2.14). Many changes in the marine fish ecosystem have been correlated with temperature changes, mostly without determining the possible cause and effect.

Sea-surface temperature (SST) changes are caused either by advection, which dominates in offshore waters, or by local exchange (heat exchange and mixing) which dominates in coastal waters and in semi-closed seas. Surface temperature can have pronounced seasonal changes (see Fig. 2.15). Its anomalies are usually smaller than 3 °C. Special consideration should usually be given to the effects of anomalies of timing of seasonal changes, temperature gradients with depth (which can depict the availability of different temperature regimes within depth), and the range of temperatures in the area of given stock distribution.

The nature of temperature (and salinity) anomalies in the open ocean can be demonstrated with long-term records which are available at a few weather ship locations in the oceans (Figs 3.10 and 3.11). These anomalies include short-term (monthly) changes as well as long-term trends. Some of the anomalies are advected downcurrent (Fig. 3.12), but considerable local changes also occur, and therefore the generalization of point observations over larger areas are questionable. Global and hemispheric SST fluctuations are unlikely. Furthermore, no real and meaningful periodicities can be observed in these records, although some longer-term irregular fluctuations are apparent.

Temperature can affect fish:

(1) as a modifier of metabolic processes (i.e. affecting food requirements and uptake and growth rate),

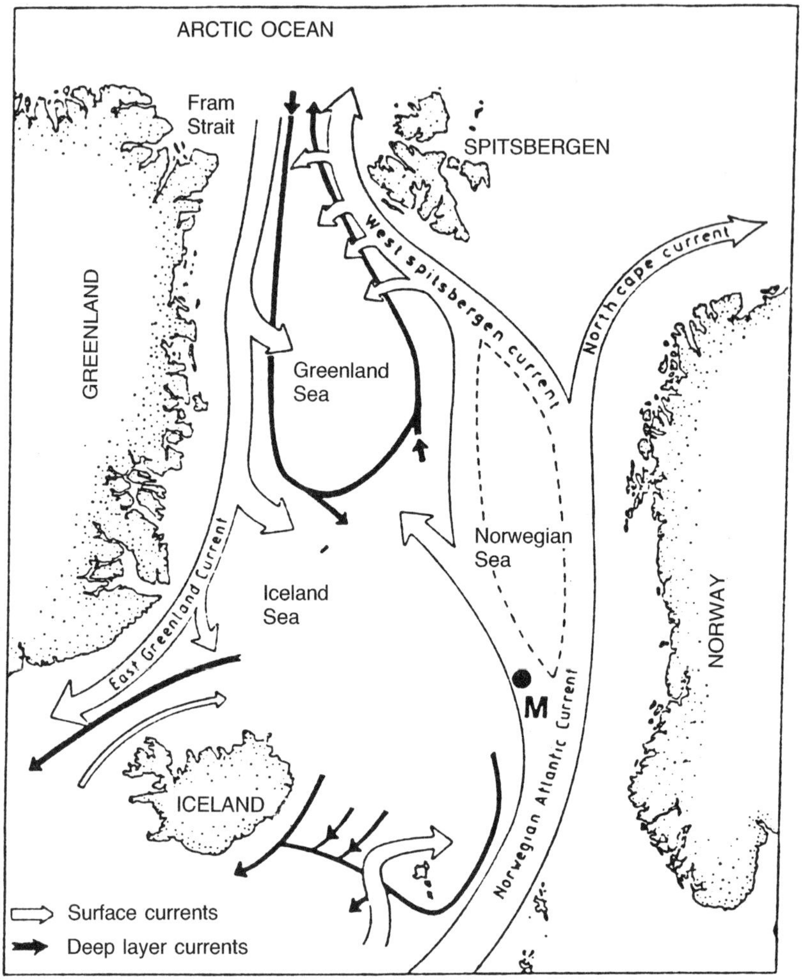

Fig. 3.10 Main current systems (schematic) in the Nordic Seas (Gammelsrød *et al.* 1991). M-Weather Ship Station *M*.

(2) as a modifier of bodily activity (re. swimming speed), and
(3) as a nervous stimulus.

How fish react to the temperature anomalies might be complex. It could be assumed that most reactions of fish species to the environmental anomalies occur on synoptic to monthly time scales. Longer term, seasonal and annual, reactions must include some integration process, such as change of feeding areas via migration and search, or some effect on growth rate, maturation, and even on recruitment.

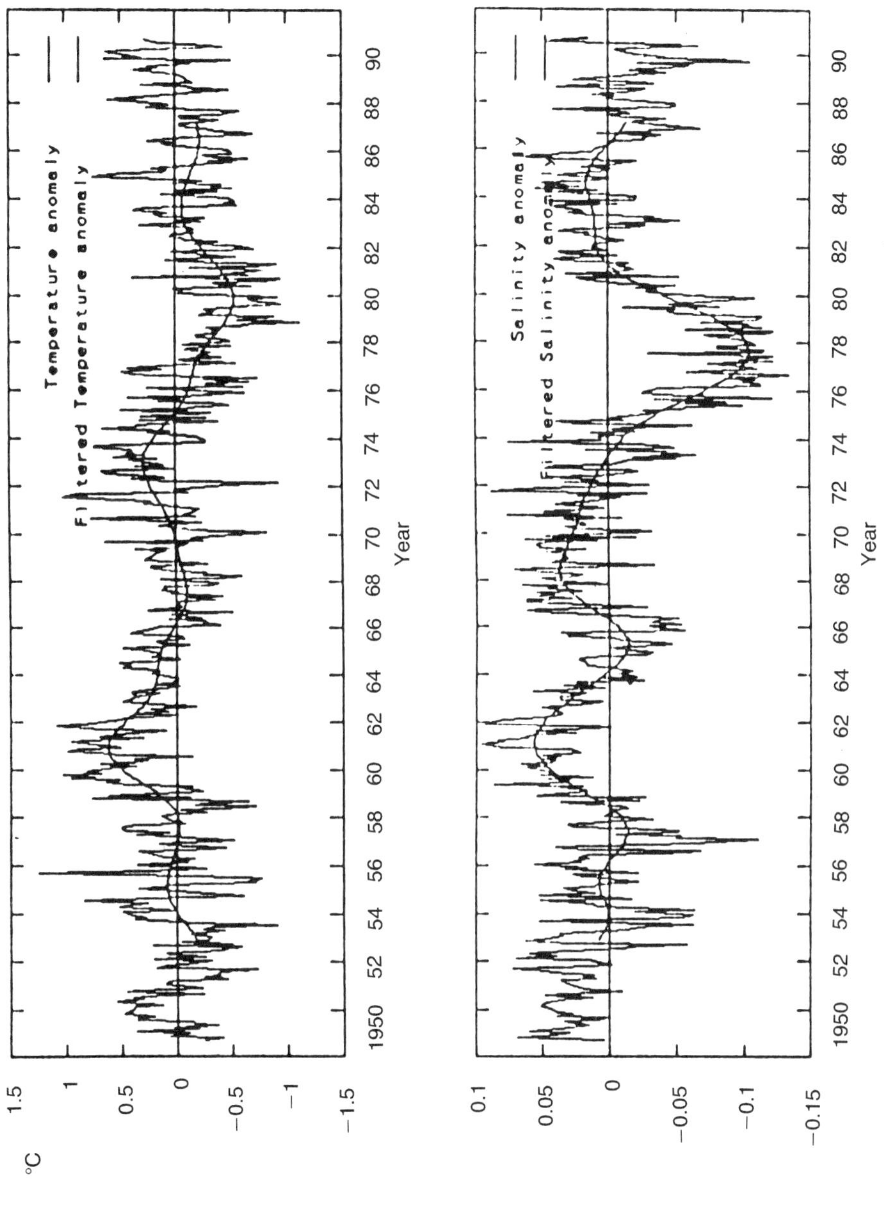

Fig. 3.11 50 m temperature and salinity anomalies at Ocean Weather Station *M* (Gammelsrød *et al.* 1991).

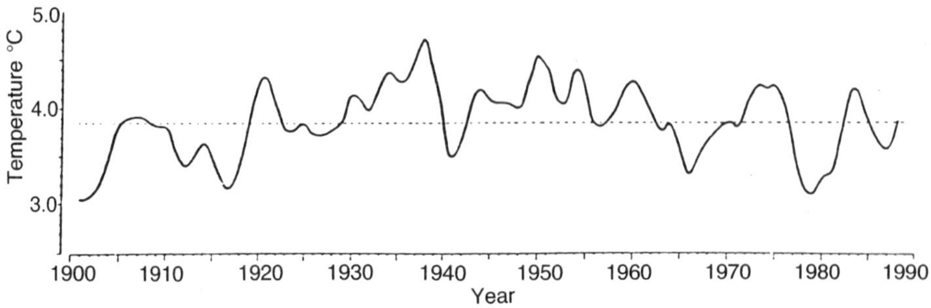

Fig. 3.12 Three years running average of yearly temperature in the Kola-section during the period 1900–1990 (Loeng *et al.* 1991).

Major changes in landings of several of the world's pelagic species have occurred (Fig. 3.13a,b). It has been postulated that large-scale climatic fluctuations might have influenced the changes of the landings of these species, and the authors of Figs 3.13a and b have indicated some climatological events on these figures. However, as stated earlier, no global and large-scale regional climatic changes in the sea which might affect the pelagic species can be identified; instead local changes are sought. Indeed, Shannon *et al.* (1984) believed that the causes of the fluctuations of pelagic stocks are related to local changes in stratification of the water column, turbulence, altered current patterns, or changes in local food supply. Some of the causes might be within the ecosystem itself through the interaction between dominant and subdominant species (Skud 1982, 1983) (see Section 6 of this chapter).

As water temperature change can be indicative of changes of wind systems, upwelling, and currents, concomitant changes to the fish ecosystem can be expected to be apparently related to temperature changes. LeClus (1990) found off Namibia that with negative SST anomaly (intensified upwelling) the spawning of pilchard and anchovy shifted northward, whereas with positive anomalies (relaxation of upwelling) spawning took place both in the north and further south. Thus temperature is unlikely to be the direct cause of changing spawning location.

Year-to-year changes occur in environmental conditions as well as in fish occurrence and behaviour. However, it is often not possible to relate these changes by cause and effect. In 1968 and 1969 extensive joint investigations by Iceland, Norway and the USSR were conducted on the distribution of herring in relation to hydrography and plankton in the Norwegian Sea in May and June (Anon 1969). In 1969 the herring followed the same migration as in the previous year in spite of slightly higher temperatures in most of the eastern Norwegian Sea and colder than normal conditions NE of Iceland. The mean migration speed was between 12 and 15 nautical miles per day, occasionally reaching 30 nautical miles per day. It was, however, slightly higher in 1969 than in 1968. Whether slightly higher temperatures influenced this speed difference can only be guessed. When herring reached an area east of Bear Island they remained nearly stationary in waters of 3 to 4 °C. The shoals remained deeper (150–300 m) in 1969 than in 1968 and were feeding on small squids. Zooplankton concentrations were slightly lower in 1969, but no relation between herring occurrence and zooplankton abundance could be established.

Nakken & Raknes (1987) confirmed the expected effect of temperature on growth of fish. They found that the growth of cod in the Barents Sea increased significantly during the period of observation, coinciding with an increase in mean temperature in the distribution areas of the fish during the same period.

Midttun *et al.* (1981) found a westward migration of cod with increasing age and with decreasing water temperature in the Barents Sea. These migrations of older cod into deeper and cooler waters are natural phenomena observed in several gadoids elsewhere. Southward and westward displacement of all age-groups of capelin in Barents Sea was also observed by Loeng *et al.* (1983). This displacement coincided with decreasing water temperatures in the region. The water temperature anomalies in the Barents Sea are mainly caused by wind anomalies (Saetersdal & Loeng 1983), causing advectional changes in the surface layers, which can also cause anomalous transport and distribution of fish larvae and plankton. Thus the apparent southward and westward movement of capelin might well be related to advection with the Barents Sea surface waters. Loeng (1989) stated that rich year-classes of cod in the Barents Sea occur only in years with relatively high temperature on the spawning grounds and in the areas of distribution of larvae and juveniles during the first half-year of their lives. He concluded that climatic fluctuations also influence the plankton production and thereby the food conditions for all plankton feeders. Temperature fluctuations may therefore indicate the variability of food supply and the availability of the latter may be as important as the direct effect of temperature on the biological conditions of fish. Both temperature and food (plankton) fluctuations may in turn be a direct effect of advection and mixing by currents.

Sockeye salmon in Bristol Bay arrive for the spawning run in a remarkably short period (about 2 weeks), with small variation in timing from year to year. McLain & Favorite (1976) found an apparent delay in timing of the run during anomalously cold oceanic conditions in the southeastern Bering Sea. Nishiyama (1977) (cited by Burgner 1980) compared the mean sea surface temperature of the Bering Sea in June and the timing of the Kuichak River sockeye runs for the years 1965–71. He concluded that the runs are earlier in warmer than in colder years. One can only hypothesize that these migration timings might result from the effect of temperature on the migration speed of sockeye.

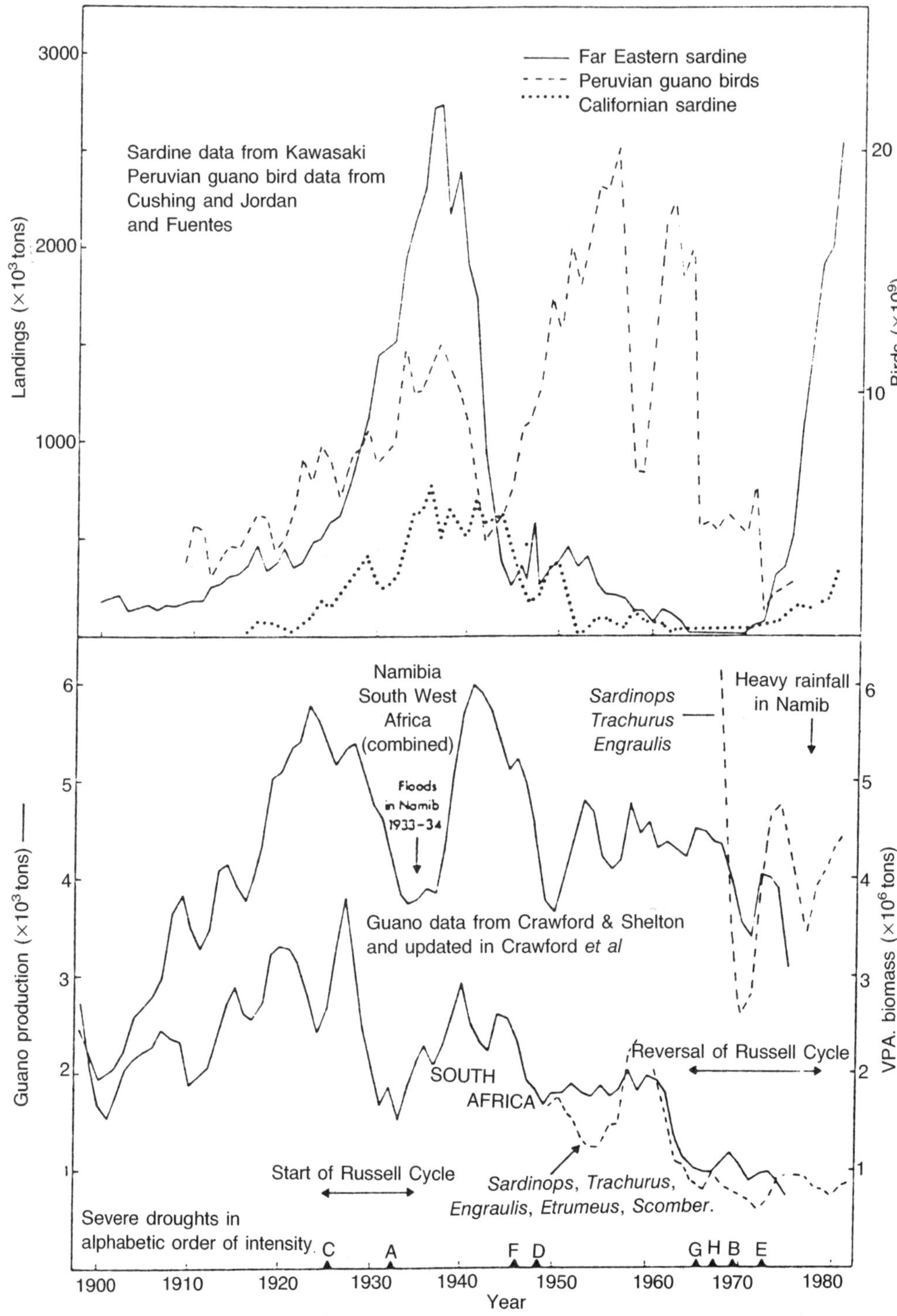

Fig. 3.13 (a) Long-term trends in the landings of Pacific sardines, the abundance of Peruvian seabirds, and guano production off South Africa and Namibia. Superimposed on the last are summations of VPA estimates of biomass for species caught by purse-seine boats (Shannon *et al.* 1984).

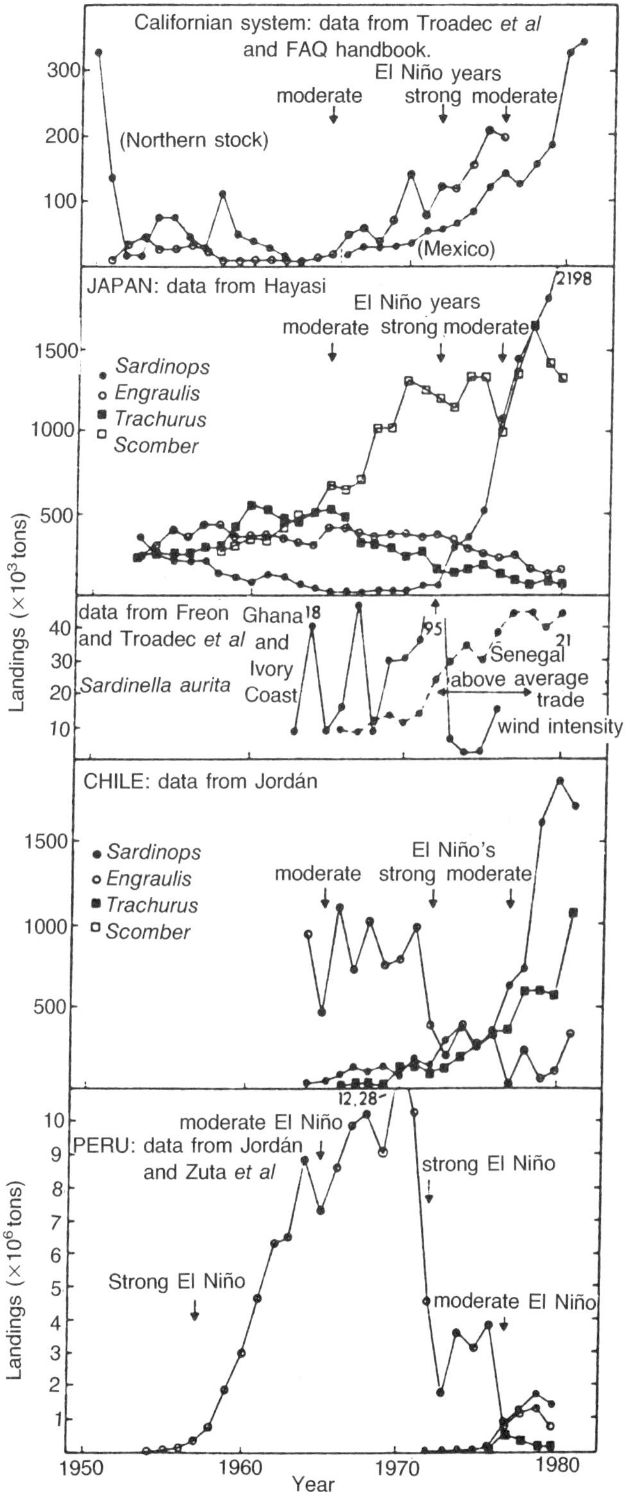

(b) The pelagic catches of some fisheries, 1950–1982 (California, Japan, Chile and Peru) (Shannon *et al*. 1984).

Grainger (1978) found that the herring abundance fluctuations off the west of Ireland have about a 10-year cycle and that winter temperature and salinity anomalies 3 to 4 years earlier showed some association to these fluctuations. He concluded that the effects on the short-term herring fluctuations are not simply temperature-dependent, but influenced by changing associated oceanographic conditions. A large part of the variation of catches could reasonably be attributed to changes in the abundance of herring off the Irish west coast, either by stock movement or by changes in stock biomass. The relationships to temperature might be entirely coincidental.

From the selected examples in this subsection one might conclude that water temperature change and/or anomaly are seldom the direct cause of changing availability and abundance of species, but might indicate that other processes which might have more direct influence on fish distribution have occurred in the region.

3.5 Sea ice, cold winters and fishery resources

Sea ice and cold winters are climatic phenomena of high latitudes and of some semi-closed seas, such as the Baltic Sea. The climatic fluctuations of ice cover might be expected to affect the fishery resources, especially their growth rates and migrations. On the other hand the high-latitude fish ecosystems and species therein are adapted to the occurrence and fluctuations of these severe conditions and most species overwinter happily under the ice. Many subsistence as well as sports fisheries operate on the ice.

Kalejs & Ojaveer (1989) found that in the Baltic Sea rich autumn herring year classes have occurred only in warm first-winter conditions (in average at 53% or less of mean ice cover). Rich year classes of sprat occur after warm winters when conditions for wintering of the warm-water zooplankton are favourable. Similarly, Sjöblom (1978) found that spring spawning herring are most abundant following the years of severe winters and that autumn spawners are most abundant during mild winters. It is postulated that the observed herring and sprat abundances are directly related to the abundance of zooplankton. Similar effects of cold conditions and ice extension on zooplankton production and ultimately on the occurrence of the Atlanto-Scandian herring north of Iceland, have been described by Jakobsson (1980).

The extent of sea ice in the ocean is determined primarily by advection and secondarily by air temperature (i.e. winds from high latitudes). Icing might also be connected with cold bottom water formation on the continental shelf (e.g. the Bering Sea). Some fisheries operate near the ice-edge, e.g. cod fisheries in Greenland and Labrador, where good catches are made. It is not certain whether the good catches are caused by aggregation of cod near the ice edge or sluggishness of cod in cold temperatures and thus easy capture, or by both effects.

Ice and cold winters affect fishing, especially coastal fishing and fishing near the ice edge, where superstructure icing might also occur. In areas outside the regular winter ice cover occurrence, cold winter might cause the migration of coastal species into deeper water.

Abnormally cold winters can cause fish kills in medium latitudes, whereas such kills are not known from high latitudes. Simpson (1953) described some such kills from the southern North Sea. During January 1947, a high pressure system over Scandinavia caused strong easterly winds over the North Sea for approximately 4 weeks. This displaced several species from their normal spawning grounds (dab, cod, flounder) and delayed hatching. Dead fish (cod, plaice, and dab) were found in several cruises. A similar event occurred in 1929.

The effects of ice cover and cold winters are a rather common phenomenon in high latitudes, where both fish and fisheries are adjusted to these conditions. Only some occurrences of exceptionally cold winters at medium latitudes are part of regional climatic fluctuations. Their effects are usually local and limited to shallower coastal areas.

3.6 Fluctuation of stocks caused by biological processes within ecosystems

Many processes within the fish ecosystem affect the fluctuation of stocks without either the influence of environment (climatic anomalies) or influence of fishing. Simple correlation studies of climatic effects on stocks usually overlook the internal processes of the ecosystem. Examples of internal fluctuations of the ecosystem without environmental effects are the cyclic fluctuations of predominant cannibalistic species, such as the walleye pollock, in which age-specific fluctuations of cannibalism cause fluctuation in abundance of stocks (Laevastu & Favorite 1988). Further examples of internal changes in the ecosystem are periodically-occurring diseases (e.g. eel pest), excessive predation by mammals, and natural changes of migration routes and changes of principal spawning areas.

Internal processes in the ecosystem which cause stock fluctuations might also be triggered by slight environmental anomalies or by fisheries and highly magnified within the ecosystem. Examples of these changes are recruitment changes and changes in dominance.

There is rarely any relationship between the spawning stock size and recruitment. Stocks have been known to recover from low levels. Examples are the Japanese sardine, North Sea herring and Icelandic capelin between 1982 and 1983 (Vilhjalmsson 1987). The changing predation conditions are believed to play an important role in recruitment control (Laevastu & Bax 1989) as the predation mortality is the largest mortality component (Table 3.3) and the greatest determinant of recruitment (Table 3.4).

Table 3.3 Mortalities in fish stocks (Laevastu and Favorite 1988) (modified)

Mortalities	Relative magnitude
Natural mortalities	
Predation	In most cases the largest component of natural mortality
Senescent or spawning stress	May be largest component of mortality in older age classes
Diseases	Little known, assumed small
Morphologically malformed larvae	Probably minor
Environmental extremes	
Cold	Occurs seldom
Oxygen (lack of)	Rare local occurrence
Pollution	Small, local (e.g. estuaries)
Fishing (mortality)	None to very large, depends on intensity of fishing

Table 3.4 Determinants of recruitment (a general summary)

Process, condition	State of knowledge
Spawning	
Size of spawning stock Egg survival and hatching Turbulence (e.g. by storms) Temperature anomalies	Great variability in S/R relation Effects known, little quantified, but usually small
Larvae (survival)	
Starvation	Might occur, but not a major factor
Predation	Evidence indicates that it might be major factor in larval mortality
Transport	Little quantified, variable in space and time
Prefishery juveniles	
Predation	Recent analyses and multispecies research show predation to be main process controlling juvenile abundance
Other mortalities	Less quantified than predation (see Table 3.3)
Emigration/immigration	Magnitude and variability little known

Considering that predation is by far the greatest component of mortality of fish and that it operates mainly on larvae and juveniles, it becomes obvious that predation will materially affect the recruitment in most species in those areas, such as the Bering Sea, in medium and high latitudes where fish diet is a considerable fraction (e.g. 22%) of the fish ecosystem diet. The predation rate depends on the relative sizes of predator and prey, their relative (size specific) abundance, and the availability of alternative prey. Thus predation is dependent on the relative density of prey items. Consequently, if recruitment were mainly affected by predation, it would vary relatively little from year to year and long-term trends would be evident when the ecosystems composition (i.e. dominance of species within it) changed.

The carrying capacity of a given ocean region with a high standing stock of fish depends not only on the production of basic food (zooplankton and benthos), but also on the availability and production of fish as food for other fish, i.e. the species composition and general trophic status of the fish ecosystem as well. If the level of a prey species such as capelin or herring in a given ecosystem is low, the predator species might suffer partial starvation and the predation pressure on their larvae and juveniles would increase through cannibalism and recruitment would be reduced. This seems to have been the case in late 1980s in the Barents Sea (Hamre 1988; C.C.E. Hopkins, pers. comm.). The recovery of predominantly piscivorous species in such a depressed fish ecosystem might be slow owing to the increased predation pressure on the larvae and juveniles of these fish. Growth rates of the piscivorous fish might also be affected.

Skud (1981) reviewed classical papers on species (anchovies, herring, sardine, mackerel) and environmental interactions causing changes in dominance. He cited alteration of dominance by pilchard and herring (Cushing & Dickson 1976), mackerel and herring (Lett *et al.* 1975) and sardine and anchovy, and discussed possible climatic and interspecies interactions causing these changes. Furthermore, Skud (1982) explored the relationships between the dominant and subdominant species in relation to possible climatic changes. A summary of Skud's conclusions is given below:

> 'Changes in abundance of dominant and subordinate fishes in response to changes in the physical environment showed that dominant species responded positively to factors that improved their survival, and subordinate species responded negatively to the same factors. The responses changed when dominance changed. This inverse relation indicates that the abundance of the subordinate species is controlled by the abundance of the dominant species.
>
> If this hierarchy is ignored, the interpretation of a correlation between the abundance of a subordinate species and an environmental factor may be erroneous because interspecific interactions may mask the actual relation. Changes in dominance explained why a species, in one area, responded positively to temperature for many years and then responded negatively, and why a species, in different geographical areas, did not respond in the

same way to temperature. Correlations that fail should not be summarily dismissed as spurious until changes in dominance are considered. Dominance should be examined when responses of either the same or closely associated species in different years or areas are compared.

The maintenance of distinct levels of abundance for two interacting species suggests that exploitation reduces the biomass of the dominant species below its equilibrium density, enabling it to respond to favorable climatic factors. If competition is reduced through a reduction in the biomass of the dominant species, the subordinate is capable of increasing through intraspecific interactions even though climatic factors may not be entirely favorable. The dependence of the subordinate abundance on that of the dominant species would explain the lack of a stock recruitment relation for certain species.

The inverse relation of the abundance of the dominant and subordinate species to physical factors in the environment and the observed changes in dominance support Nicholson's thesis that climatic factors can affect the abundance of a species but do not govern its absolute density and that the mechanism controlling abundance within a population is intraspecific competition. In addition to its bearing on ecological concepts of population control, the effects of dominance have particular significance to problems in fishery management.'

Chapter 4
Effects of Weather on Fish Availability and Fishing Operations

Throughout the ages most fishermen have observed weather and associated various weather conditions with the availability of fish in given specific locations and depths, and have adjusted their fishing strategies according to the weather to achieve better catches. This, usually local and species- and weather-specific knowledge on weather–fish relations, has been based on the personal observation and experience of fishermen and on knowledge passed from father to son.

Little attention has been paid to explaining why and how weather affects fish, and few of the experiences have been written down. However, lately some research has been conducted in some countries, notably in Norway and in Scotland, on fish behaviour in relation to gear. This research is aimed at making gear more efficient and in some instances more selective to meet the requirements of fisheries regulations.

Examples of weather-dependent, simple catching strategies are the setting of nets and long-lines to a specific depth or in a specific direction in relation to wind direction and speed. The direction and depth of trawling are likewise often adjusted to wind and current. Other simple examples of weather-related fishing phenomena are higher catches of salmon and eels in fyke-nets during storms and lower catches in drift nets during moonlit nights.

Marine fishermen have had little help from scientists in studying and explaining phenomena which would improve fishing. However, for sports fishermen in rivers and lakes several books of hints exist. Carruthers (1966) pointed out that as fisheries research is almost entirely state-funded its concern has to be with food from the sea as a national requirement and not in helping the fishing industry. Most often the fishery authorities are concerned to limit fishing and to make it inefficient to conserve the resources and to maintain good catches in the future.

A quotation by Carruthers (1966) of Dr W.M. Chapman describes well the relations between the fishing industry and fisheries and marine scientists:

> 'One thing the fishing industry and scientific community appear to be in substantial agreement upon is that when scientists do science and fishermen fish a good deal is learned and a lot of fish is caught, but when fishermen assay science and scientists attempt to show fishermen where and how to fish all hands should better have stayed in bed.'

In this section only a few general notes on the effects of weather on fishing and fish can be given. Most of the practical knowledge of fish behaviour and availability to gear (often termed by scientists anecdotal knowledge) is location, species, season, and gear specific. It could be desirable to collect the experiences of experienced fishermen and skippers, re-evaluate and record these in writing for benefit of future fishermen as well as for fisheries research. It could indeed be expected that the observations by fishermen would provide valuable suggestions of subjects for further research by fisheries scientists. Much of the knowledge of fish behaviour and availability to capture in relation to weather and ocean conditions is proprietary and personal. Wise use of this knowledge separates the good skipper or fisherman from the mediocre one. Good fishing is not good luck, but good application of knowledge.

Weather, through its integrated effects on the ocean, affects the availability and catchability of mainly pelagic and semi-pelagic species. To enhance the national fisheries, fisheries forecasting has become a separate practice and science in some principal fishing countries, and manuals have been written on it (Bocharov 1990).

Often fisheries forecasts are extrapolations of historical knowledge of fish migration, spawning and other behaviour, and of recent catch statistics in relation to, and as influenced by, observed weather-influenced ocean features. These include location and movement of current boundaries deduced from surface temperature and its gradient, upwelling, sighting of birds, and water colour as an indication of plankton and micronecton (Burbank & Douglass 1969). The environmental features are easier to observe than fish and will thus lead fishermen to locations where good catches are expected. The forecasts are usually presented to skippers as fish location maps.

The knowledge of the relationships between weather and fish availability (occurrence) has become less important in the last few decades, because most offshore vessels use echosounders and sonars to find fishable aggregations of fish. Small manuals for successful tactical operation of the sonic fish finding equipment have been prepared by equipment manufacturers (e.g. Simrad 1964).

Ocean and fisheries forecasting services, in some respects comparable to meteorological services, exist only in a few countries, including Japan and the former USSR. Attempts to establish ocean services in other countries have been made, but these have largely foundered. The reasons for failure have been manifold; attempts by meteorologists to design the service like the meteorological service, but lacking knowledge of fisheries and oceanography and of the purpose and use of ocean services; lack of knowledgeable personnel; lack of practical knowledge and experiences; and, above all, lack of feedback from fisheries.

4.1 Wind, waves and mixed layer depth in relation to depth behaviour of fish

Fishing operations are mostly affected by the effect of winds on the ocean surface, the most apparent effect being the waves. The operation of ship and gear is mainly affected by higher seas. In higher latitudes the icing of the superstructure is one of the specific hazards for fishing vessels, which is also related to winds and to air and water temperatures.

Surface winds could affect fish through their effects on the ocean, mainly by wave action and associated turbulence and oscillatory movements within depth. Mixing by wind waves, generated by stronger winds, also determines the mixed layer depth during the warmer seasons, spring and summer. Wind also generates surface currents and their convergences/divergences (see Chapter 3).

Fishermen observe the present and near past winds and utilize past experiences of the relationship between wind and fish availability for their tactical fishing decisions. Scientific studies of wind in relation to fish behaviour are few, and most deal with wind as a climatic element affecting fish. However, there are two noteworthy studies of the effects of wind on catches, one at Hamburg (Walden & Schubert 1965) and another at Lowestoft. Two papers from Lowestoft (Harden Jones & Scholes 1980, Scholes 1982) dealt with trawling for bottom fish, mainly plaice and cod. These studies were prompted by the belief of Lowestoft fishermen (trawlers) that fewer fish are caught with winds between northwest and northeast ('starvation winds').

Harden Jones & Scholes (1980) found that catches of plaice in the southern North Sea were low during northerly winds all the year round, whereas catches of cod were higher with these winds, mainly in autumn and winter. The authors of this study pointed out that a relationship between the catches and winds might arise if a given wind direction prevails during a spawning aggregation season of the species. To guard against this spurious relationship, consequential trawl hauls, between which a wind shift occurred, were examined, and it was found that the results were consistent with a causal relationship between wind and catch.

The real causal relationship between wind and plaice catches was believed to have been found in wave-caused oscillation near the bottom, which stirs up sand and causes high turbidity near the bottom. Northerly winds were mostly associated with longer swell, which causes deeper oscillatory movements. During the heavy swell the plaice either burrowed into the sand or moved into deeper water. It was reported that Grimsby fishermen successfully work deeper holes after gales, where they make heavy catches before the fish disperse. A further study by Scholes (1982) partly confirmed the results of the earlier study, but showed that the results are quite variable and not always conclusive.

The distribution of many pelagic fish is known to respond to wave action, fish moving into deeper layers when heavy sea and swell is present. There is also a relationship between waves and mixed layer depth during spring and summer, and the depth distribution of many pelagic fish and their diurnal vertical movements are related to thermocline (see Section 4.2).

Nearly every fisherman is convinced that some relationships exist between wind direction and availability/occurrence of fish. Unfortunately, for various reasons this varied knowledge is not scientifically tested. This knowledge pertains mostly to local (coastal) conditions and cannot be generalized. It is reported (Hodgson 1957, in Walden & Schubert 1965) that no good catches of herring can be expected off the coast of East Anglia with easterly winds. Westerly winds are supposed to drive herring out from Norwegian fjords, whereas easterly winds are related to good catches there. West winds are also favourable for good catches off the Danish west coast.

Most fish orientate to a current by heading into it. Thus the movement of herring is usually upwind, i.e. heading into the wind current. The investigations off the coast of East Anglia (Hodgson 1957) further demonstrated the upwind movement and gave the following results. When the wind turned from the northeast to the southwest or southeast the herring moved close to coasts and catches were good. When the wind changed from S to E, the area of main distribution of herring moved further offshore, the shoals became smaller, and catches decreased. The same situation arose when the wind changed from northwest to southeast.

Few wind–demersal fish relationships have also been reported. Mohr (1964) reported that in the Lofoten area the saithe dispersed from fishing grounds, probably by dispersing into a water mass above the bottom when the wind shifted to a northerly direction. This dispersal sometimes occurred before the wind shifted. Mohr believed that this dispersal was caused by internal waves, excited by travelling surface pressure systems, rather than by local winds.

Walden & Schubert (1965) examined more than 45 000 wind and herring catch records of German herring luggers from different fishing grounds in the North Sea between 1955 and 1960. Catches by trawl were examined separately from catches by drift-nets. Wind directions and magnitudes were correlated with each in only a few cases. For instance, in the Flemish Bay above average catches occurred with SW and WSW winds but also with weak ENE winds. These investigations were concerned with the offshore rather than with the coastal fishery.

Many of the wind–fish relations reported in the literature pertain to larval recruitment. The main reason for this is that fisheries research is concerned with resource (stock) fluctuations rather than with problems pertinent to fishing, such as fish availability. Carruthers (1938) linked wind conditions off

England with herring catches and recruitment. When winds are from the Channel, herring spawning products drift to a favourable environment and later become available to the Ostend fishery. Chase (1955) found that the brood strength of Georges Bank haddock can be predicted from the occurrence of offshore winds during the pelagic phase of eggs and larvae. Similarly Koslow *et al.* (1987) found that the year-class strength of haddock and cod in the NW Atlantic is somewhat associated with offshore winds. Neither reported finding has been tested in subsequent available data. Lasker (1981) has described how year-class strength of the northern anchovy may be dependent on storms, their direction and duration, by affecting food concentration in the near-surface layers. However, the underlying hypothesis of this cause has now been challenged by results from mariculture research, notably from Norway, showing high survival of larvae with relatively low food concentrations.

Many early German investigators, notably Rodewald (in Laevastu & Hayes 1981) have related fisheries to wind anomalies. In a document from the Federal Republic of Germany (1960) to CMM of WMO (Commission of Maritime Meteorology of the World Meteorological Organization) it was pointed out that catches of redfish off the coast of southern Labrador (Sundall and Ritu Banks) were very good with pronounced on-shore anomaly winds. It was further pointed out that the cooling or warming of the Barents Sea was related to wind anomalies which in turn were related to migrations and the availability of cod, haddock, and saithe in this area. This observation has been amply substantiated by several later Norwegian investigations.

Upwelling off the coast is related to wind conditions, which in turn may be related to fish availability. Schneider & Methven (1988) found that spawning of capelin on beaches in eastern Newfoundland often occurs after wind-driven upwelling ceases and during periods of light wave action on beaches (see also Chapter 2 section 7 and Chapter 3 section 4).

4.2 Changes in diurnal behaviour of fish in relation to weather

Many pelagic fish undertake diurnal vertical migrations. These migrations might be affected by weather elements, e.g. wind (waves) and light (cloud cover). In normal migrations the fish rise from near the bottom or near the thermocline to near the surface layers during dusk, disperse there and later sink and compact (shoaling) into deeper or near-bottom layers during dawn.

The diurnal vertical migrations of albacore tuna correspond to vertical migrations of zooplankton and small pelagic fish (Burbank & Douglass 1969). These deep vertical migrations might be independent of weather and light, but might be dependent on season (deeper during winter) and on local water types and their stratification, especially near current boundaries. The vertical migrations of salmon are strictly diurnal, rising to near the surface during evening, where they are accessible to gill-nets. However, in heavy seas salmon tend to remain deeper.

In summary, the vertical diurnal migrations can be complex indeed and very variable from species to species, from season to season and from area to area. The vertical behaviour of a given species changes throughout its life cycle. The vertical migrations are also affected by the vertical distribution of proper food organisms and by the temperature structure with depth (see Chapter 2 section 5).

The last mentioned condition leads to a number of questions, especially if evaluating reports where temperature fluctuations are ascribed as direct cause for (climatic) change in fish ecosystem. Why should relatively minor surface temperature anomalies affect horizontal migrations of pelagic fish if they can move only a few metres or few tens of metres vertically through the thermocline to find a desired temperature range, rather than to migrate tens or hundreds of kilometres horizontally to find the same temperature? How would fish know in which direction to move horizontally if they would be in search of a preferred temperature, when they can probe locally within the depth of any desired temperature?

4.3 Storms and coastal fisheries, and effects of sound on fish behaviour

Several reactions of fish to storms have been reported, for some of which a hypothetical cause can only be guessed at. In some cases fish react before the arrival of the storms. Whether this reaction is caused by early arriving swell or by currents caused by atmospheric pressure changes is not known. Porpoises in Mexican waters have been reported as leaving exposed areas several hours before the onset of bad weather (Bernard 1973). Harden Jones & Scholes (1980) quote an observation by Dunn that herring and pilchards off Cornwall coast move from shallow to deep water before the onset of a storm some ten or more hours before any of its violence could reach their locality. These fish reactions before storms could be caused by either internal waves excited by moving surface pressure systems or changes of currents caused by the same phenomenon. Reaction to noise, caused by breaking swell waves on the coast, cannot be ruled out either, as this noise travels in water and is within the hearing frequency range of fish (50–700 Hz).

Blindheim *et al.* (1981) found that the fluctuations of temperature along the Norwegian coast are of an advective nature. These fluctuations are usually driven by a wind current component, which would also advect pelagic species. Thus the availability of fish at any given location in coastal areas may be affected by wind, as experienced fishermen know, and this might be caused at least in part by wind-driven currents. However, research on this subject is very limited. Furthermore, any knowledge of winds and fish occurrence is location-, species-, and gear-specific. Thus the collection and utilization of this knowledge will remain fishermen's privileged information.

Some coastal areas are affected by major oceanic fronts (e.g. Southeast Iceland, Stefánsson 1969). Considerable fluctuations in the positions of these frontal zones are caused by winds by means of wind-driven surface currents, which in turn affect the availability of fish, the distribution of which is related to these fronts.

The fluctuations in coastal upwelling are usually effected by longer lasting prevalence of given wind and falls within the realm of climatic fluctuations rather than weather phenomena.

It is well known that fish have good hearing in the low frequency range and react to noises. (For information about fish sounds and hearing see Laevastu & Hela 1970). Ship propeller noises as well as vibration noises of ropes are well within the hearing frequency of fish, which might react to them in various ways.

The reactions of fish to ships' noises (mainly propeller noises) have received attention from some researchers (e.g. K. Olsen, Tromsø). Figure 4.1 shows

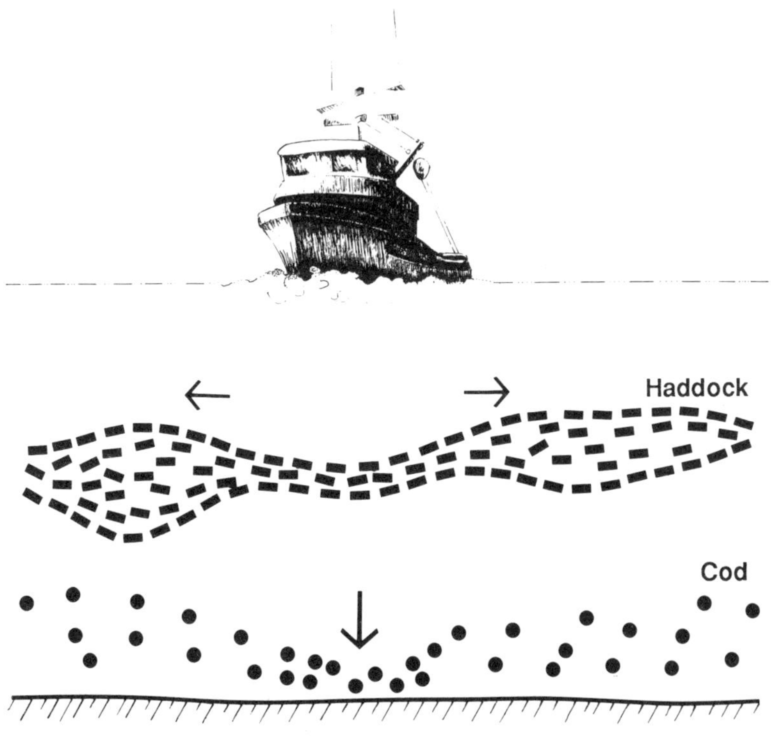

Fig. 4.1 Schematic representation of the reaction of haddock and cod shoals to ships' noises.

schematically the response of haddock in mid water and cod near the bottom to the noises of ships, including echo sounders) (E. Ona and S. Olsen, personal communications). Haddock shoals in mid water tend to disperse to both sides of the ship, whereas cod packs close to the bottom. If this behaviour of cod occurred in all places and seasons, then a noisy ship would catch with the trawl better than a quiet ship; the reverse would be true for haddock. It has been claimed that ships' noises break up large pelagic shoals of fish and might affect their migration courses. Furthermore it should be noted that trawl wires also cause considerable noise, and that the Danish seine works largely on acoustic principles.

4.4 Fish behaviour in currents, and the effects of currents on long-lining, nets, and trawling

Fish sense currents with the rheotactic organ located on the lateral line. Generally fish head into the current even when they let themselves be carried with it. The swimming speed of the fish depends on their size, and is affected by temperature, being slower in lower temperatures. The seasonal and life-cycle migration (e.g. spawning migration) speeds are usually the maximum sustainable speeds (Table 4.1). Fish eggs, larvae and small juveniles are carried with currents and dispersed by them.

Harden Jones *et al.* (1976) believed that most fish movements and migrations are related to currents. Young stages drift passively with currents from the spawning grounds to the nursery grounds. Migrations to the spawning grounds are contra-natant movements. Consequently distributions of particular stocks appear to be contained within regional circulation systems (gyrals).

The strongest current component in coastal areas is the tidal current. Fish react and respond to it and utilize it for migrations. Greer Walker *et al.* (1978) studied tagged plaice movements in the North Sea and found that plaice utilize tidal current for selective tidal stream transport. Fish usually came off the bottom to mid-water at slack water, moved downstream with

Table 4.1 Speeds of migrating fish (Favorite and Laevastu 1981)

Species	Speed (km d^{-1})
Sole	7–16
Plaice	1–7
Herring	4–30
Salmon (sockeye)	54
Salmon (chum)	48
Halibut	6
Herring	25
Yellowfin Sole	3–7

tide, and returned to the bottom at the next slack tide (Fig. 4.2). Fish moved little on the bottom with an opposing tide. These researchers also found that the depth below the surface at which the fish swims was affected by weather, especially with wind force above 7.

Inshore-offshore movement of shrimp and elvers seems to utilize similar selective tidal stream transport. It can also be postulated that spent fish, which are weak after spawning, are transported away from spawning grounds by currents.

Japanese fisheries scientists as well as fishermen have long recognized that pelagic fish tend to aggregate at current boundaries, where good catches are made (Burbank & Douglass 1969). The reasons for this might be threefold;

(1) food supply (micronekton) accumulates at current convergences;
(2) the current boundary acts as an environmental boundary; and
(3) migrating fish dynamically aggregate at current boundaries.

Most offshore fishing can be profitable only on fish aggregations. If predictions of fish aggregations in space and time were possible, fleets could be directed to these aggregations, thus reducing the search time for fishable aggregations and increasing the economic returns from investments in fisheries. Fish aggregation by artificial means (e.g. light and sound) is used in some fisheries (e.g. night-light purse-seining for anchovies in the Mediterranean).

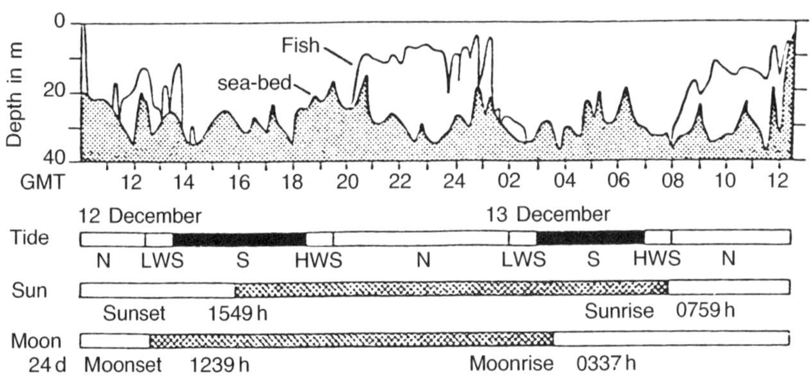

Wind	WSW	WSW	WSW	WSW	WSW	WSW	SSW	WSW	SSW	SSW	SSW	SSW	SSW	SSW
Force	4	5	4	4	3	3	3	4	4	5	5	5	5	5
Sea	1	1	1	1	1	1	1	1	1	2	2	2	2	2
Swell	1	1	1	1	1	1	1	1	1	1	1	1	1	1

Fig. 4.2 The depth of a tagged plaice in relation to the direction of the tide and other environmental factors (Greer Walker *et al.* 1978).

Fish aggregations are caused by innate, species-specific behaviour, influenced by several physico-chemical and biological environmental conditions, but finally effected by fish migrations. The dominant behavioural aspect of aggregation is migration to spawning grounds and spawning. Temperature anomalies can cause changes in the timing of peak spawning and dislocation of spawning from traditional spawning grounds. Whether wind-driven surface currents affect spawning migrations and timing of spawning is, however, not known with certainty.

Outside the spawning season, fish aggregations can be caused by feeding and migration interactions. For example, fish shoals slow the speed of migration in areas of abundant food. Furthermore, as the speed of migration is affected by temperature and by currents, it can be expected that fish aggregations can occur at water type and current boundaries (fronts). Fish may also concentrate at seasonally preferred depths along migration routes. Multiple factor interactions, such as vertical migrations stimulated by light and

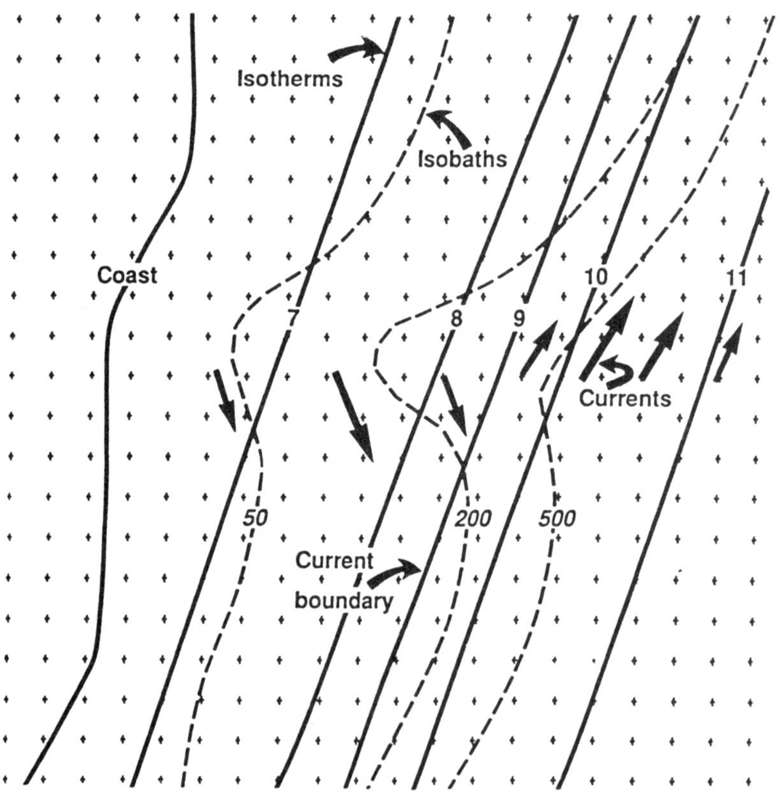

Fig. 4.3 A hypothetical configuration of bathymetry, temperature and currents used for simulation of fish migrations and aggregations in Fig. 4.4.

wave actions, can enhance aggregations in specific locations on the continental shelf. All these mechanisms of aggregation can be simulated by numerical methods which give plausible locations of probable fish aggregations.

Examples of the results of a numerical model simulating fish migration in a hypothetical area with given depth, temperature, and currents are given in Figs 4.3 and 4.4. Figure 4.3 shows the bathymetry, currents, and temperature distribution for the hypothetical numerical model of fish migration. An initial circular distribution of fish (Fig. 4.4) is made to migrate northward with a prescribed basic migration speed which is affected by depth, temperature, and currents. Although each environmental condition separately affects the distribution of migrating fish, the combined effects create aggregation at the continental slope, one side of the current and temperature front after 13 days of migration, similar to the aggregations known to occur in frontal zones in the sea.

Unfortunately, synoptic information on currents is not available to fisher-

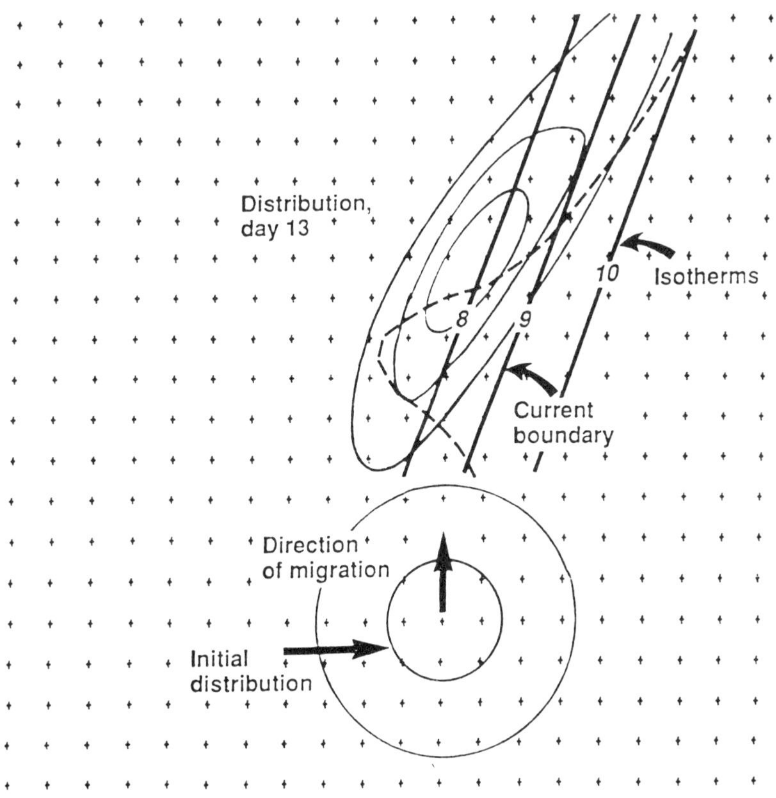

Fig. 4.4 Initial distribution of fish and distribution of the same fish at the current boundary after 13 days of migration.

men. Spot measurements of currents are possible with moored current meters but a synoptic current chart can be computed only by numerical means. The fisherman can use only indirect observations which indicate the presence of a current and water type boundary. Surface current boundaries are recognized by water colour change, by strong temperature gradients, and by accumulation of flotsam. When a larger eddy is cut off at the current boundary, the fish contained in it are expected to remain in it for some time, unless they migrate (e.g. in a search for food or an innate seasonal migration).

The migration of large pelagic fish, such as salmon and tuna, and their response to currents is largely hypothetical. These fish migrate in the surface layers over the wide ocean and often in circular routes. It is presumed that they use currents for their orientation, as some gill-net experiences indicate. Furthermore, different stocks seem to remain in defined parts of current systems (Fig. 4.5). Salmon species in the North Pacific start their homeward migration over wide areas, but arrive at their rivers in a remarkably short time. Temperature anomalies are known to affect the homing migration speed (McLain & Favorite 1976).

The effects of currents on gill-nets, whether drifting or anchored in shallower water, are well known to fishermen. That this knowledgc has trickled down to fisheries biologists (among others) who use gill-nets for resource assessment is not always apparent, and many myths about gill-nets for squids (the so-called 'curtains of death') have reached the environmentalists. Obviously the use of gill-nets and the effects of environmental processes on their catching properties varies considerably. There are few scientific summary works on this subject available in literature.

The effects of currents on trawl catches vary with the target species and its behaviour in currents as well as with the trawling speed, which is mainly a function of the horsepower of the vessel. The skipper can seldom know in which direction the tidal current reigns, so he makes his decision on the direction of tow according to the wind and wave conditions.

The effects of currents on long-lining was studied by Olsen & Laevastu (1983). Fish are attracted to bait mainly by olfactory stimuli. Past studies show that most olfactory attracted fish swim against the current, and that currents near the bottom distribute the smell from the baits. The emission of attractive olfactory stimulants from the bait decreases rapidly with time so that within an hour about 40% of the total soluble proteins, which are the main stimulants, may have dissipated. The interactions between changing leaching rate of smell, current speed and direction in relation to the line, especially during the first few hours after setting, and mixing of the smell fields from adjacent baits greatly influence the catch rates with long-lines. Therefore the direction of the setting of long-lines in relation to the prevailing near-bottom current, and change of the tidal current, should be considered in long-line fishing (Olsen & Laevastu 1983).

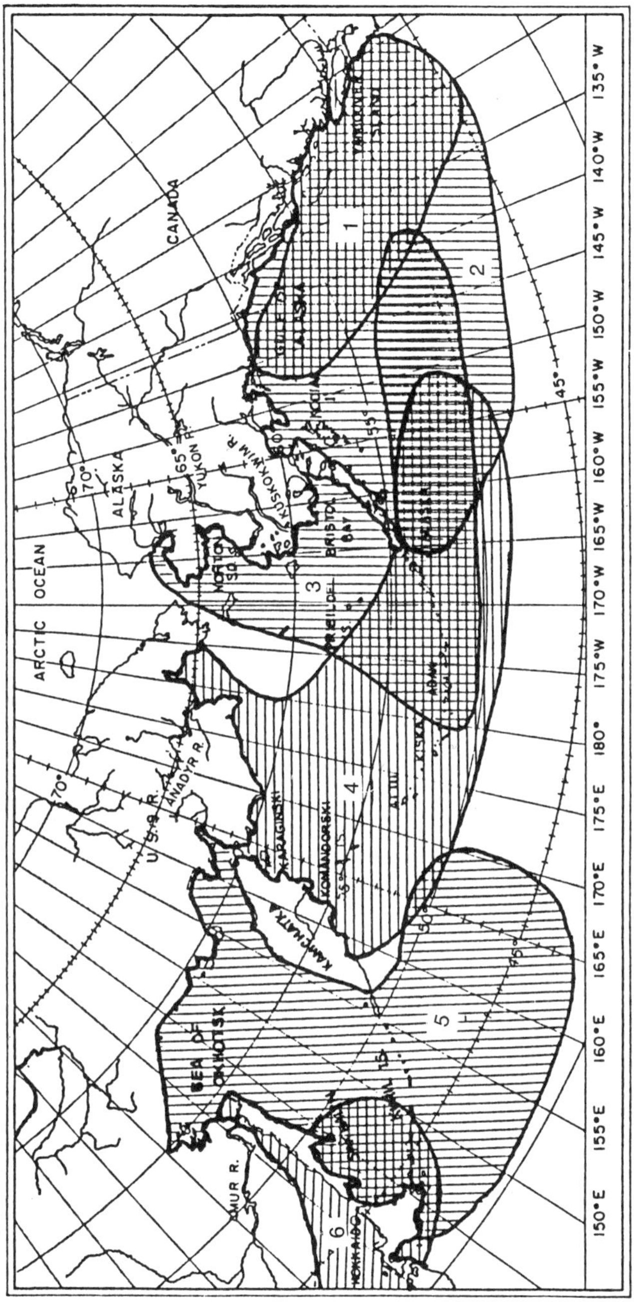

Fig. 4.5 The primary ocean distribution of pink salmon originating in different geographical regions: (1) Washington and British Columbia; (2) SE, central, SW Alaska (3) N side of the Alaska Peninsula to Kotzebue Sd; (4) E Kamchatka to Anadyr Bay; (5) W Kamchatka, N Okhotsk Sea coast, E Sakhalin Is, Kurile Is, Hokkaido Is; (6) W Sakhalin Is, Amur River, Primore Coast (Burgner 1980).

4.5 Sea surface temperature distribution as an indicator of the effects of weather on fish

Surface water temperature is an integrator of the action of several past weather elements on the sea surface layers and is also an indicator of past processes and present conditions in the sea. The temperature changes are mainly caused by three processes, advection (by currents), mixing (e.g. deepening of mixed layer depth), and local heat exchange processes between the air and the sea.

Elizarov (1980) believed that the rate of change of an environmental property such as temperature, and the corresponding intensity of the process causing the change, might be a more important factor in changing fish behaviour (activity and adaptation) than the actual state of the environmental property. This hypothesis might be true, as most changes of fish distribution occur during the processes which affect the environment most rapidly.

The horizontal surface temperature gradient is a good indicator of the location of current boundaries, which are either areas of fish aggregation or, in some cases, diffuse fish distribution boundaries. Laurs & Lynn (1977) pointed out that forage fish in a temperature gradient (front) may be a far more important factor in fish aggregation than the response of tuna to a temperature gradient. Laurs *et al.* (1977) have ascertained by tracking sonically tagged tuna that albacore show a tendency to congregate in the vicinity of coastal upwelling fronts, presumably to feed. They move away from the immediate area when upwelling ceases and the upwelling front is no longer present at the surface.

Recruitment is not directly related to fish availability and fishing. However, recruitment fluctuations have often been related to temperature fluctuations and to changes in wind-driven surface currents. A discussion of recruitment in relation to climatic fluctuations is found in Chapter 5 section 2. Only a few examples of temperature, current and recruitment relationships are given below.

Martin & Kohler (1965) found that recruitment of cod in the southern Grand Banks and Nova Scotia area was negatively correlated with sea surface temperature during the first year of life. Whether this correlation can be translated as increased southward transport of the cold Labrador Current is uncertain. On the other hand, Tereshenko (1980) believed that meteorological conditions (winds) largely determine the surface current transport of cod eggs and larvae from the Lofoten area to the Barents Sea. This northward transport also causes temperature increase in the southern Barents Sea and might be an explanation why good year classes of cod arise with warming in this area (Saetersdal & Loeng 1983).

In the 1950s and 1960s considerable attention was paid to fish location by temperature regimens (e.g. Dietrich *et al.* 1957). Rogalla & Sahrhage (1960) studied the distribution of herring in the North Sea in relation to temperature but found no strict temperature–herring relationship, although better catches were made in the tongues of warmer Atlantic water. The colder, lower salinity water of the Skagerrak was avoided by herring in the early summer.

The relationship between temperature and fish occurrence can often be brought about by the affiliation of given fish species to a water type, i.e. fish remaining in a given water mass and being advected with it.

Thermometric methods for fish finding have less importance nowadays as fish location is done with electronic equipment (echo sounders and sonars).

4.6 Weather and fishing (summary)

Research on the availability and aggregation of fish and fish behaviour in response to environmental conditions and processes in the oceans has been scant. Reasons for this are many, none of them being fully defensible from the fishing industry's point of view. The emphasis in fisheries research has been on resource assessment and lately on fisheries management. For practical fishermen the fish location and fish aggregation are of more immediate interest than the resource size. However, the fishing industry at large is concerned about the present state of the fishery resource and its availability in the future. It is to be hoped that the objectives of fisheries research will change; there are some signs that the research of fish reaction to gear has intensified in the last decade.

Most of the existing and applicable knowledge of the effect of weather on fish comes from observations by fishermen. Much of this knowledge depends on the location and type of fishery. The importance of the knowledge of location of fish aggregations and availability to gear in relation to weather and ocean conditions has decreased somewhat for offshore fishing, since echo sounders and sonars have become commonly used for fish finding.

The greatest and quasi-immediate effect of the weather on fishing and fish behaviour is caused by action of surface wind on the ocean. These wind effects are temporary, local and progressive in time and space, as the weather systems are progressive in space and time. On the other hand, the integrated effect of weather, i.e. the climate, determines the hydroclime and affects to some extent the fishery resources, their distribution and abundance.

Possession of some knowledge of the sea–air interaction processes and their resulting effects on the ocean (Chapter 3) will permit fishing skippers and other fishermen to interpret the observed effects of weather on the ocean and to use this knowledge to improve their catches.

The most important weather element at sea is surface wind. Its main effects on the ocean are the generation of surface waves and currents in the surface layers. Both effects cause turbulent mixing in the sea and affect the ocean thermal structure with depth. Many fish react to this turbulence as well as to current, but this reaction varies from species to species and in space and time. Wind direction and its change can determine the availability of pelagic fish in specific locations in coastal waters. In offshore areas no wind direction–fish availability relation has been found. Storms can repel fish from near coastal areas. In offshore areas storms affect the depth distribution of pelagic fish and in a few areas demersal fish also, the latter being apparently affected by a long swell.

Research in the past decade has demonstrated the effect of ship noises (propeller sounds and possibly echo sounders) on fish shoals, which consists mainly of ship-avoidance reactions. This knowledge has some application on practical fishing, e.g. the need for quiet propellers.

The knowledge of prevailing currents in given locations and at given times can have applications in long-lining and gill-netting. The knowledge of the positions of more persistent current boundaries can be used for the location of aggregations of migrating fish.

Sea surface temperature is useful as an indicator of current boundaries, and with some precautions also of surface water types. However, temperature has otherwise little use for fish finding or as an indicator of fish behaviour. Thermometric fishing has demonstrated little practical applicability.

Chapter 5
Climates and Fishing in the Past and Comparison with Other Factors

5.1 Separation of the effects of fishing from the effects of climatic changes on stocks

The fluctuations of catches and landings of any species are caused by numerous factors, such as changes in fishing fleets and technological developments of gear, changes in market demands and prices, and changes in stock sizes and/or availability. The stock size and availability of fish to the fishery are in turn affected by past fishing intensity, variations in recruitment, changes in processes in the fish ecosystem (e.g. predation, competition) and various changes in environment on different space and time scales.

Man affects the fish stocks, not only by fishing, but also by causing pollution in coastal waters and eutrophication in semi-closed seas and by influencing the abundance of the natural top predators of fish, the marine mammals. The effects of pollution have not always been negative, but in some areas positive through eutrophication. Elmgren (1989) found that in the 20th century the fish catches in the Baltic Sea have increased more than tenfold owing to increased production of fish food caused by eutrophication, increased fishing effort, and considerable decrease of fish-eating Baltic seals through hunting and pollution.

In order to evaluate quantitatively the possible effects of climatic changes on stocks and catches, we must separate and evaluate all the factors affecting the catch and/or stocks. Seldom is a single factor responsible for the changes in stock size and/or catches; many factors may operate simultaneously and in different directions. It is, therefore, difficult to determine the contribution of any one factor to the temporal and spatial changes in fish ecosystem. Unfortunately many preconceived assumptions, interpretations and rationalizations have been used to interpret these changes in the past, and data are usually not available for re-evaluation of these reports.

In this section the difficulties in separating the various effects of environmental (climatic) changes and fishery and other factors on fish stock and landing fluctuations are briefly reviewed. The difficulties in separating the effects of fishing on a stock and its collapse from the effects of possible minor environmental changes have been colourfully described by Radovich (1982), who pointed out several faulty interpretations of basic data and 'politicking' in the management of the resources.

Earlier Bell & Pruter (1958) found that many climate temperature–fish productivity (landings) relationships do not give sufficient provision for changes in fishing effort to explain climate–fish relationships because the effects of other variables such as economic conditions, changes in fishing practices and the extent of the removals by man have often not been accounted for.

Interactions between fish availability, migrations, and abundance

The availability of a given species on traditional fishing grounds can alter owing to changes of fish behaviour and annual migration routes, thus affecting the catches of given fleets. These alterations in availability can occur without changes in the stock abundance but might be related to them. Most often it is not possible to find any apparent cause for such alteration, although there might have been minor environmental changes, such as changes of net (or resultant) currents within a given season which might have affected the migration of shoals. The changes of migration routes of some major pelagic species, such as Atlanto-Scandian herring, are described in Chapter 5 section 2 and Chapter 7 section 2.

One of the complex changes of the German haddock fishery at the turn of the century has been described by Lundbeck (1962) (see Fig. 1.1). In earlier years the haddock fishery took place in winter and spring for spawning haddock on the Great Fisher Bank, whereas later it became a summer and autumn fishery on feeding fish. Part of the haddock stock which spawned on the bank migrated to the German coast each year. However, when the haddock withdrew to the more northern part of the North Sea, their migration no longer extended to the German Bight, but fish aggregated for feeding on the Great Fisher and Dogger Banks. Because of the changes in haddock migrations and distribution, the German steamers turned to the middle North Sea grounds, where they fished on younger fish, and consequently the percentage of small haddock suddenly increased in catches, which was mistakenly taken as a sign of overfishing. However, a natural narrowing of the area of distribution of haddock with an intensifying fishery resulted in the decline of the haddock fishery in the North Sea at the turn of the century. Lundbeck (1962) concluded that

> '... the question arises whether natural causes or overfishing (intensive fishing) are active (in changing the stock size), and obviously cannot in all cases be answered in a distinctive way, but the interaction of both is possibly far more common than it is thought until now.'

Hempel (1978a) reviewed the changes in fisheries and fish stocks in the North Sea in the 1960s and 1970s and tried to explain them by natural and man-made effects (see also Chapter 5 section 2 and Chapter 7). He pointed out that until the mid-1970s each stock was considered separately when evaluating the effects of climate on year-class strength or the effects of heavy fishing on the stock. This single stock concept did not take into account the interactions between the various fish stocks and the complex interactions with the environment. Hempel (1978b) concluded that natural environmental factors might have been more important than fishing and fisheries management in causing the fluctuations of stocks but this conclusion cannot be proved. He listed as environmental causes of stock fluctuations the predation on larvae and timely availability of suitable food for earlier life stages.

Lluch-Belda *et al.* (1989) concluded that the large fluctuations of the November–May catch of both the Monterey and crinuda sardines were due to their availability to the fishing fleet, and that availability was largely influenced by the ocean climate, especially by cool years with northerly winds and intensive upwelling or by warm years that might result from an El Niño event. Again, this statement cannot be tested, nor can the causal mechanisms be explained.

The disappearance of herring from the Bohuslän archipelago shortly before the turn of the century gave, indirectly, the stimulus to organize the International Council for the Exploration of the Sea. In the last 90 years many hypotheses have risen about this disappearance of abundant herring, which have never returned in the same abundance as before to the Bohuslän coast. None of the hypotheses have been definitely and convincingly approved with data. The same can be said about the latest plausible hypothesis of Lindquist (1983), who suggested that the periodic appearance of herring in the Bohuslän archipelago might have been caused by two independent mechanisms, many strong year classes and major hydrographic changes, mainly caused by changes in westerly winds over the North Sea. Herring might, however, have altered their migration routes and feeding and spawning areas for reasons unknown, as has been the case with Atlanto-Scandian and North Sea herrings (see Chapter 7 section 2).

The interaction between climatic fluctuations and changes in fish stocks can be complex indeed, involving several species and fisheries. This can be briefly illustrated with recent changes in the Barents Sea. The Norwegian Sea and Barents Sea contain several important commercial species, which migrate according to their lifecycle needs (Fig. 5.1). A kind of crisis situation has risen in the Barents Sea (1987 onwards), whereby all commercially important stocks have decreased drastically. Hamre (1988) believed that this crisis in the Barents Sea was linked to an unbalanced state of the predator–prey relationship which developed after the herring stock was fished out in the late 1960s. Herring and cod were the largest stocks in the area and their recruitment was apparently determined by common

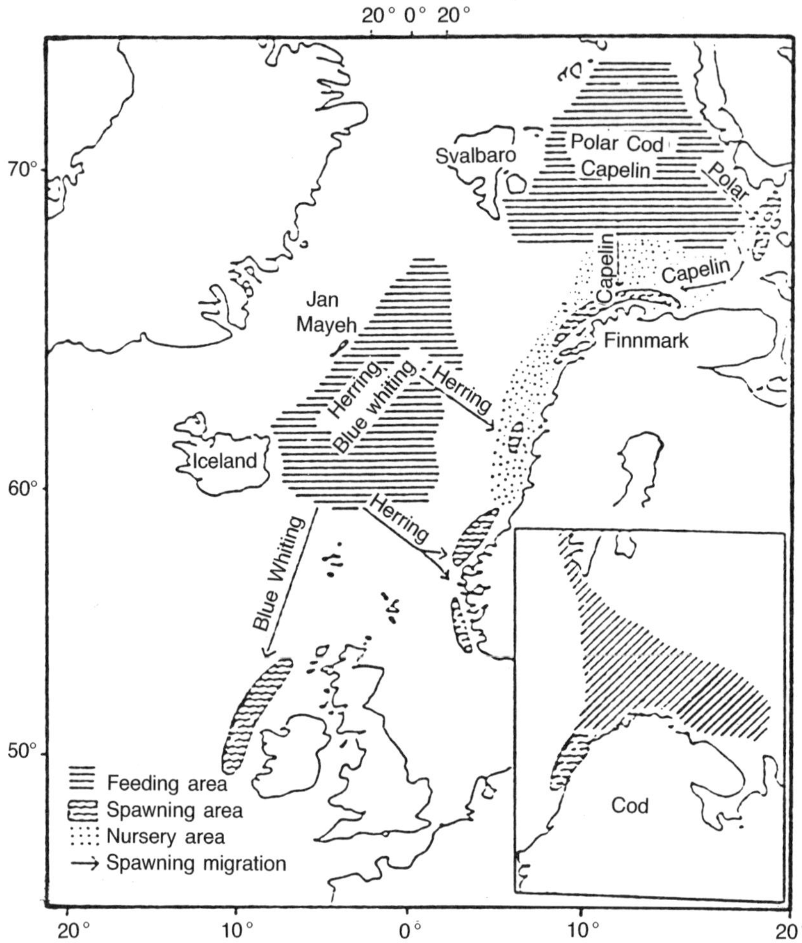

Fig. 5.1 The general distribution of herring, capelin and cod in the Norwegian Sea–Barents Sea ecosystem (Hamre 1988).

environmental factors, whereby the recruitment success of capelin was inversely related to the abundance of the juvenile herring stock. He furthermore believed that the releasing factor for the crisis was a shift from cold to warmer climate in the early 1980s which favoured the recruitment of herring and cod. Very strong year-classes of cod were recruited, but the spawning stock of herring was low. The rapidly growing cod stock grazed down the plankton feeders in the area between 1983 and 1986 and starved cod, sea birds and seals have appeared on Norwegian coast since 1986. Heavy capelin fishing also accelerated the process.

Some of the examples mentioned earlier show that fish availability to the fishery will change when migration routes or feeding areas change. Alterations to the migration route might be caused by environmental changes, especially in wind-driven current, whereas feeding areas might change owing to changing availability of prey (zooplankton and/or small forage fish).

The availability of a given species does not necessarily change in the central areas of its main distribution when the stock size changes, but the fringe areas of stock distribution might be subjected to drastic changes of availability. It is known from many pelagic and some semi-demersal stocks that their area of distribution extends when stock size increases, and new spawning grounds are established. When the stock size decreases, the distribution shrinks to the central areas, and the fringe stocklets disappear.

It could be pointed out that a spurious correlation between recruitment and fish abundance and environmental factors can arise owing to errors in the natural mortality rate (M) used in virtual population analysis (VPA) for stock assessment (Lapointe & Peterman 1991). The natural mortality rate is not known with any reasonable accuracy for any stocks.

Effects of fishing on stocks

Effects of fishing on stocks can be computed, assuming that recruitment remains constant in each year. Figure 5.2 gives the changes in relative age

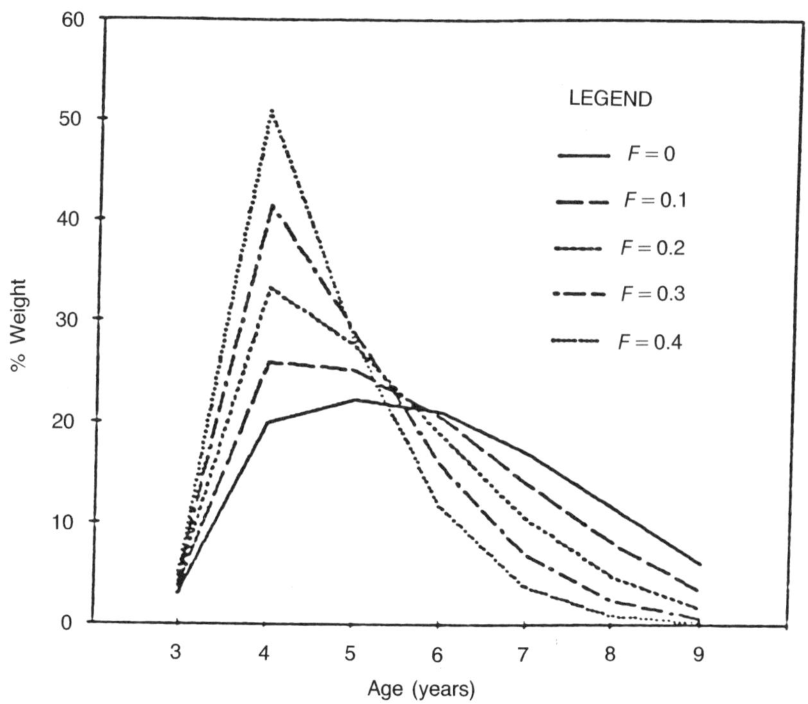

Fig. 5.2 Percentage of biomass in catches of Pacific cod in different age classes with different fishing mortality and constant recruitment.

composition of catches, in weight, with changing fishing intensity, showing the increase in the fraction of recruits (age-4 fish) in catches with increasing fishing mortality (F). Figure 5.3 shows the relative weight of the fish biomass by age left in the sea with a fishing intensity of 20% by numbers from each age class. The reason for differences between the catches of long-line and trawl is that long-line caught fish are older and fully recruited to the fishery 1 year later than trawl caught fish. Figure 5.4 shows the reduction in the biomass of fish left in the sea with different amounts of fish removed each year (80, 160 and 240 kg per year from 1000 kg biomass). If the recruitment remains constant and there is no relationship between spawning stock and recruitment (see Chapter 5 section 2) an equilibrium between removal and recruitment is reached. This takes 3 to 6 years in cod. If recruitment was affected by heavy fishing and if the market would accept smaller fish, the yield from a given stock would be a function of fishing intensity and a lower limit of gear selectivity, and we cannot find an acceptable definition of maximum sustainable yield from a given stock.

Obviously catch per unit effort (CPUE) changes with increasing fishing effort in some species, but not necessarily in shoaling species. However, CPUE is very difficult to define and measure, as it depends on a multitude of factors and varies with gear as well as targeting practices. Furthermore, the catchability coefficient is a variable which is negatively related to the fish

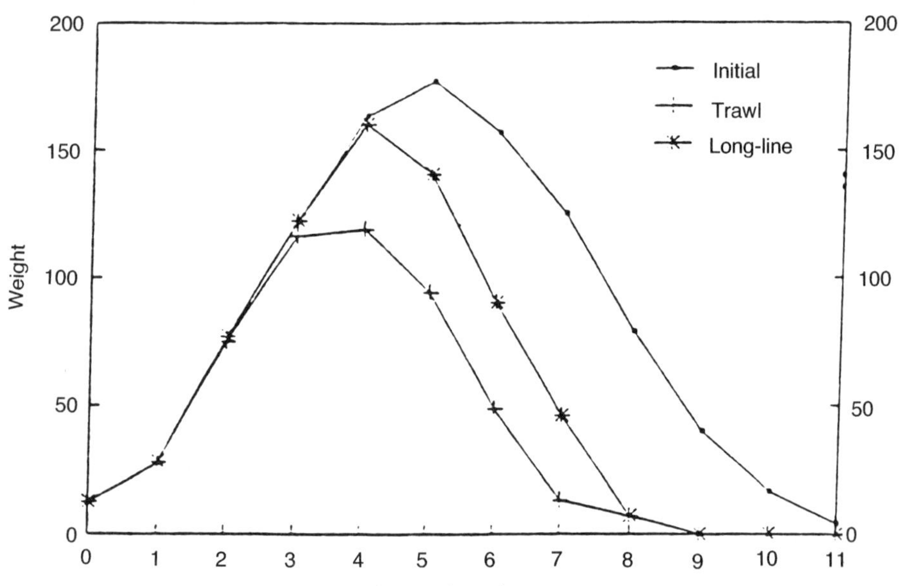

Fig. 5.3 Weight of fish of different age groups in the sea, initially and after four years of trawling or long-lining ($F = 0.2$).

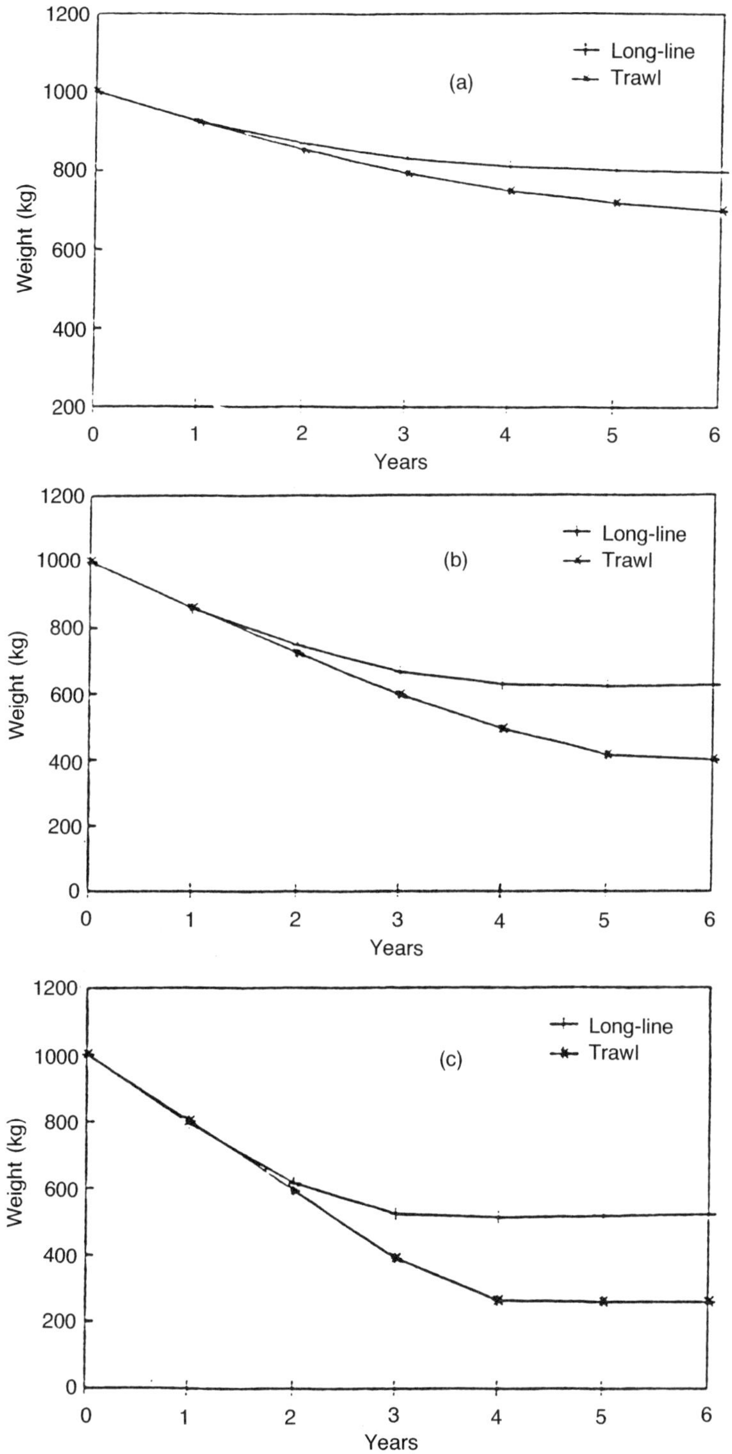

Fig. 5.4 Reduction of biomass (from original 1000 kg) of long-lining and trawling at three different catch levels, (a) 80 kg, (b) 160 kg, (c) 240 kg.

population size (Radovich 1976) and a given fishing effort may take an increasingly larger proportion of the fish population as the population declines.

A disturbing myth has existed in fisheries science, the Maximum Sustainable Yield, which has been difficult to kill, despite the excellent epitaph by Larkin (1977), who also pointed out the difficulties with fisheries statistics:

> '. . . it is also true that they (statistics of fisheries) are still incomplete and riddled with guesses, inadvertent errors, omissions and even, perhaps, some perjuries. They are generally, as a statistician would say, more precise than accurate, and that's saying something when you bear in mind that the imprecision of fisheries statistics is notorious'.

Fishing will accelerate the changes in target species biomass, especially its decline, but it cannot be entirely responsible for its large fluctuations. If a fishery is seasonal and subject to fluctuations in the availability of fish, and if the fleets have a limited sphere of action, the landings can fluctuate without changes in the total biomass of the species (Belveze & Erzini 1983).

Some fish stocks tolerate relatively heavy fishing, and stock abundance can rise in spite of intensification of fishing. Lundbeck (1962) pointed out that sole is one such species. Some heavily overfished stocks can also recover rapidly if there are 2 or more years of good recruitment in spite of a small spawning stock. Icelandic summer-spawning herring had such a rapid recovery in the mid-1970s (Jakobsson 1980).

Heavy fishing can also increase the stock size of a species which is dominant in the region and is highly cannibalistic. In this case fishing removes the older, larger and more cannibalistic specimens, thus relieving predation pressure on juveniles and thereby increasing recruitment to the exploitable stock. This is the case with the walleye pollock in the Bering Sea. Daan (1976) showed that cannibalism of their progeny by large cod in the North Sea can also affect recruitment. Besides the cannibalism, the predation by cod on sole and herring in the North Sea might affect their recruitment as well.

Fluctuations of landings and their causes

Changes in landings have often been used as the basis of investigation of the effects of climatic fluctuations on marine fish stocks as well as an indicator of stock size. However, landings data have severe shortcomings for these purposes as they are affected by numerous factors besides the abundance of fish. First of all, landings statistics are not always accurate. Caddy & Griffiths (1990) pointed out that quality control review of national fisheries statistical systems are seriously needed in many countries, and the data base shows many inadequacies.

Holden (1978) reviewed the long-term changes in landings of fish from the North Sea, and he found that besides the changes in reporting patterns (e.g. from gutted weight to nominal whole weight) and in the number of nations reporting, there were other factors causing the changes in landings. These factors were alteration of fishing patterns that lead to changes in fishing mortality rates and the patterns of recruitment to fisheries, changes in the availability of species to fleets, and changes in recruitment and growth rates. All these can be caused by either fishery or other natural factors, such as fluctuations in the hydroclime. In addition technological factors (e.g. introduction of power block) and economic factors (market demands, industrial fishery), might have affected the landings.

Increased recruitment in the 1960s accounted for the increased landings of cod, plaice and haddock, whereas market demand largely controlled the landings of whiting from the North Sea. Furthermore Holden (1978) found that there appeared to have been a northward shift in the centre of population of several species in the early 1960s, which increased their availability to capture.

There has been a considerable increase in fish landings from the Mediterranean and Black Seas from the mid-1970s to the mid-1980s (Fig. 5.5). Caddy & Griffiths (1990) suggested several plausible causes for this increase, improved statistics, technical improvements in fleets, increased intensity of fishing and increase in productivity due to eutrophication. The increase in landings has not been uniform in all areas in the Mediterranean region. It is highest in the Black Sea and Sea of Marmara, which is partly because of increased eutrophication (see Section 6.2). In contrast there has been a decrease in landings from the Levant region, which formerly received most of its nutrients from run-off from the Nile.

Saturation of small boats and/or the limited capacity of the fleets do limit the catches and landings during high abundance of species (Belveze & Erzini 1983), and so landings do not normally indicate abundance of stocks in the past when the fleet capacity was limited.

The landings of some species are subject to seasonal changes. Furthermore, if one species is not available, targeting might change to other species. The landings of sardines off the Asturias coast have experienced large fluctuations (Villegas & Lopez-Areta 1986). Before 1979 the landings were mostly below 2000 t, but rose in 1984 to 5500 t. When the landings of sardines were low, those of horse mackerel were high. Until 1965 sardines were caught in summer, from 1979 to 1982 mainly in autumn, and since 1982 mainly in winter.

Changes in composition of landings and north–south shifts of catches have occurred in the Benguela Current region (Figs 5.6 and 5.7). It is not apparent whether the alteration of species has resulted from exploitation or from changes in environmental conditions (Crawford *et al.* 1987), although

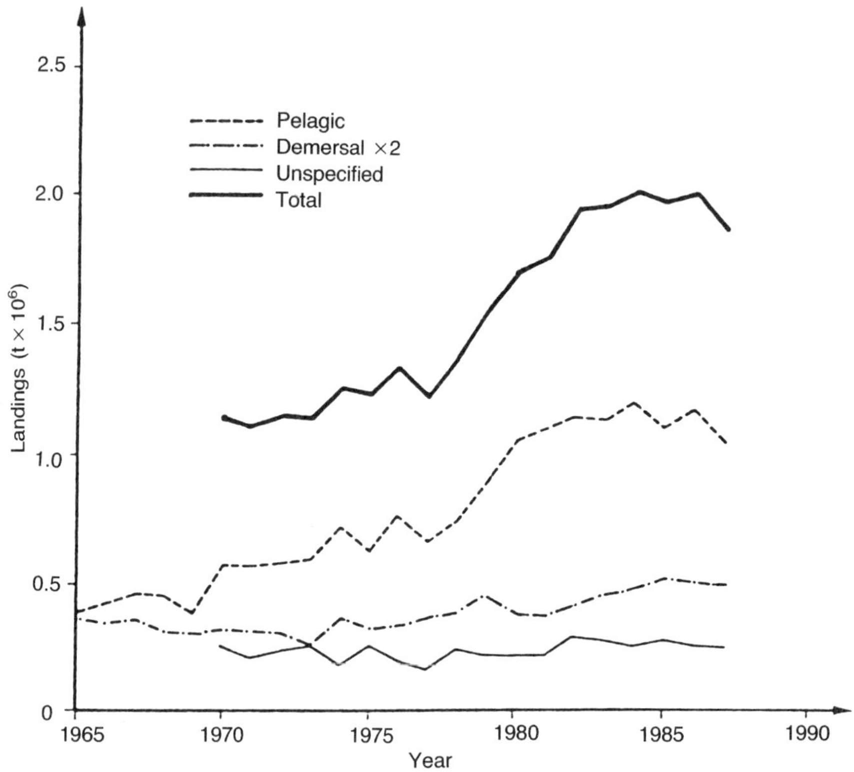

Fig. 5.5 Trends in Mediterranean and Black Sea fish landings from FAO statistical sources. (Demersal landings are doubled for ease of illustration.) (Caddy & Griffiths 1990).

in some instances resource depletions have followed immediately upon the taking of exceptionally large catches. In other instances an environmental influence, not specified by the authors, appears the more likely cause.

Considering the above, it can be concluded that landing statistics do not necessarily reflect the abundance and availability of fish stocks, nor can any simple conclusions be drawn on possible effects of climatic fluctuations using landing statistics alone. In the last decade most of the landings reflect fisheries management and other regulations and restrictions to fisheries and cannot, therefore, be used to draw any conclusions about the status of stocks.

Effects of climatic fluctuations on fishing operations

Fishing is subjected to weather conditions, especially to winds and waves, and to icing at higher latitudes. Storms prevent fishing, especially with

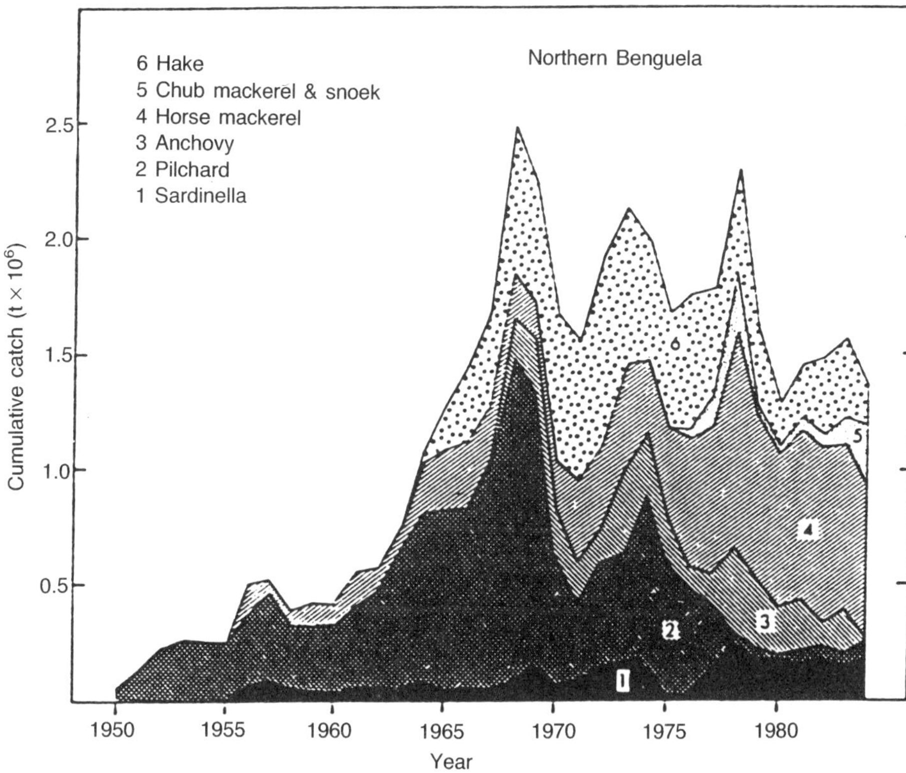

Fig. 5.6 Cumulative catches of major species groups in the northern Benguela region, 1950–84 (Crawford *et al.* 1987).

smaller vessels. Thus landings can be influenced by the frequency of storms and variations in storm tracks from year to year (see Figs 2.2 and 2.3).

New, profitable fishing grounds might be discovered in an environmentally difficult area, such as the Arctic, where superstructure icing might be frequent, or in the tropics, where refrigeration and air-conditioning of ships needs to be improved. Furthermore, if fishing regulations in a given ground prevent further catching, fleets switch to other grounds where different gear might be needed. This switching of fishing grounds often requires technical changes in fleets and/or improvements in the seaworthiness of the vessels. These technical changes and switching of fishing grounds are reflected in catches and landings. Catches from distant fishing grounds might also change owing to economic factors, such as the availability of more profitable fishing on nearer grounds.

Drift ice in high latitudes, for example in the Barents Sea and off Greenland, and solid ice cover in the Baltic can make fishing difficult or prevent it. The

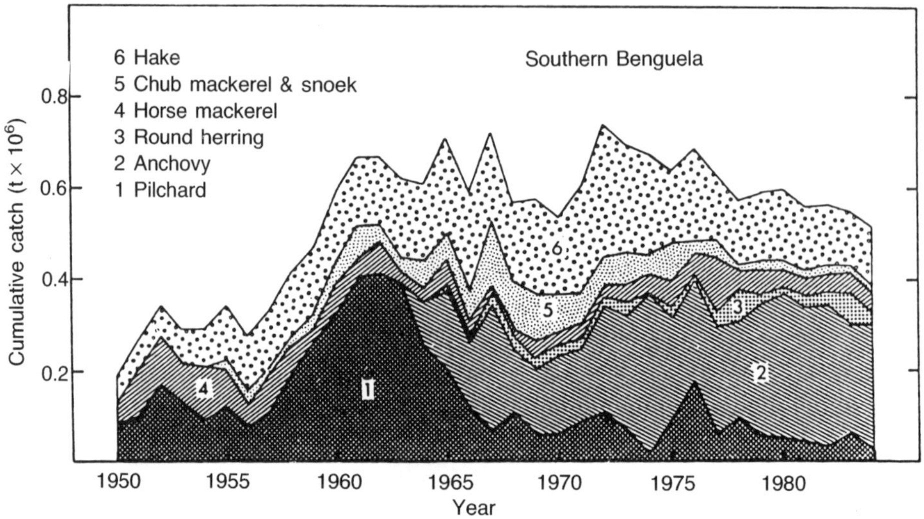

Fig. 5.7 Cumulative catches of major species groups in the southern Benguela region, 1950–84 (Crawford *et al.* 1987).

extent of ice is a climatic variable and might be thought to affect catches and landings. However, the direct effects of climatic ice cover variations on landings are relatively small, as fishing during the ice season is normally limited and/or vessels work ice-free grounds.

Relationships between coastal rainfall and fish catches have been found in lower latitudes (Fig. 5.8). Whether rain affects the fishing actively in these regions is uncertain. However, the rainfall is, mostly, related to changes in wind systems which might affect fishing as well as the occurrence of fish.

Effects of economics on fluctuations of landings

Many economic factors affect the landings of fish. The effects can vary from species to species. Holden (1978) believed that whiting markets, and thus landings in Scotland, were restricted by demands. Furthermore he found that changes in landings of some species, e.g. sprat and horse mackerel, from the North Sea were affected by the expanding fish-meal industry. The expanding industrial fishery also caused a big increase in the landing of mackerel, which was accelerated by the introduction of the power-block in high-seas purse-seining.

On the other hand, limitations of vessels (e.g. lack of onboard preservation) can restrict the sphere of action of the fleets which cannot follow migrating resource (e.g. sardines off Morocco). Furthermore the capacities of processing plants on shore can limit the landings.

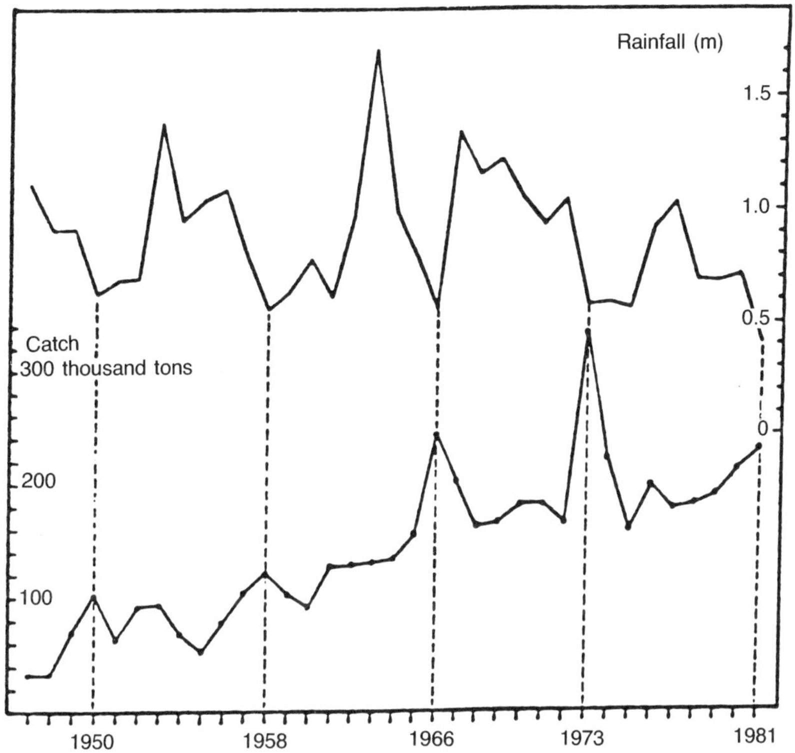

Fig. 5.8 Coastal rainfall (m) and catches of small coastal pelagic fish (t × 10^3) in zone A from 1947 to 1981 (Belveze & Erzini 1983).

An increase of coastal human population has occurred in many areas, e.g. the Mediterranean. This process has major implications for fishermen and the fishing industry as well as for fishery resources and their development, as markets of fresh fish near coasts enlarge and more people engage in coastal fishing. Coastal mariculture developments create new demands for fish food in the form of forage fish.

Various international fish trade fluctuations and import–export controls also exercise considerable influence on fish landings by changing demands.

5.2 Past fluctuations of neritic stocks and possible effects of climate

In the past, neritic fish stocks, such as sardines and herrings, have shown large fluctuations of abundance (see Fig. 3.12). As these fish inhabit the surface layers of the ocean and as any climatic change is expected to affect the environmental conditions of these layers, it has been postulated that climatic fluctuations have affected the abundance of those neritic stocks.

The effects of climate and hydroclime fluctuations might affect several aspects of the life cycle of the pelagic species. These include recruitment, migrations and distributions, through surface current changes, and the availability of food, and consequently growth, through changes in plankton abundance. Thus an examination of the possible effects of climatic fluctuations on these aspects of life cycles and processes is called for.

Recruitment fluctuations

It can be expected that the variation in recruitment, which largely influences the stock fluctuations of short-lived pelagic species, might be influenced by environmental changes, though in a complex way. It is necessary to review and understand the processes which affect recruitment before the search for connections between the environmental changes and recruitment. It is also necessary to decide whether the available data merit a search for a relationship between environmental anomalies and stock fluctuations caused by fluctuations of recruitment.

Recruitment is defined here as the quantity (number and/or biomass) of fish of a given age becoming fully available to prevailing and traditional fishing gear for the species. Thus, recruitment is determined by the numbers of young fish surviving until they are fully recruited to fishery.

Before an age-class is fully recruited to a fishery, it is affected by many processes. First, spawning is a time- and space-dependent process. Spawning periods of most high latitude fishes may last several months, there might be several spawning areas for a given stock, and the intensity of spawning in any one of these areas may vary from year to year. Eggs and larvae of many species are pelagic and so are subject to transport and dispersal by currents, which vary in space and time. This transport might take the larvae into areas where their survival, in relation to food availability and the presence of predators, can vary, and where the settling of demersal post-larval juveniles might be hindered. The growth of larvae and juveniles might be influenced by availability of proper food. The juveniles might be dispersed over large areas of ocean, and they are subject to predation which is intensive but variable in space and time.

Already before the turn of this century thoughts were expressed that the size of exploitable stock might be determined by the amount of spawn produced and the survival of the early hatched larvae. Since then numerous more or less hypothetical stock–recruitment curves have been proposed. With more data available now, we can recognize that no predictable stock–recruitment relationships exist in most marine fish (see Fig. 5.9). Only at very low spawning stock size might the numbers of eggs produced limit the size of recruitment. However, large recruitment has been known to result from a small spawning stock and small recruitment from a large spawning

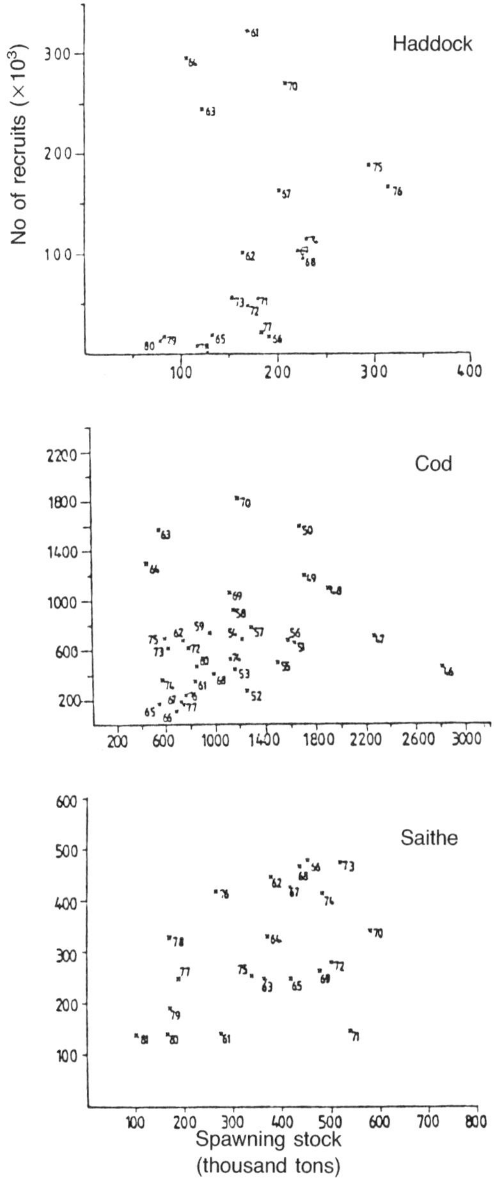

Fig. 5.9 Number of recruits (at age 3 for cod and haddock, at age 1 for saithe) against spawning stock biomass in the Barents Sea. All estimates by Virtual Population Analysis (Bergstad *et al.* 1985).

stock. It might be added that starvation of the spawning stock might affect the spawner–recruit relationship (O. Nakken, pers. comm.).

Environmental factors and processes may affect fish eggs in various ways. The spawning time may be affected by past temperature anomalies in the spawning areas through the known physiological effects of temperature on growth and on maturation of the eggs. The atmospheric environment, surface winds and their effects on the ocean, can affect the eggs and larvae in two ways, through extreme turbulence in the water column and through surface currents which transport and disperse the eggs and larvae. That heavy turbulence caused by storms and breaking waves can damage and kill near-hatching eggs and newly-hatched larvae was clearly demonstrated by Rollefsen (1932).

It has been thought in the past that newly-hatched larvae would need proper food in proper concentrations for good survival. However, the latest work in Flødevigen in Norway has shown that these hypotheses are not well founded. Very good larval survival has been achieved in Flødevigen with relatively low concentrations of naturally occurring food in sea water (Øiestad 1983).

The fish eggs and larvae may be transported and dispersed long distances and over large areas by currents. In most areas the surface currents vary considerably in time depending on the variations of surface winds. This transport can enhance as well as decrease the larval survival by transporting the larvae into areas favourable for survival as regards food and predators, or into unfavourable areas. In flatfishes larval dispersal by currents might bring part of the brood over deep water, where the metamorphosed larvae do not find a habitable bottom when settling. Usually we have neither prior nor subsequent knowledge of the favourability of these areas of destination, nor do we know the origin of the spawnings from which good survival might result.

In short-lived species the earlier larval recruitment might have some effect on the recruitment to the exploitable stock, whereas in species which are fully recruited to the fishery at age 3 or later, predation on juveniles might play an important role. Hempel (1978a) found that no convincing argument had been developed to relate the recent (mid-1970s) cluster of exceptionally good year-classes in the North Sea fish stocks to environmental changes. He believed that better recruitment might have come about because more food for the larvae was available through a higher production of small zooplankton, better seasonal and geographical match of food production and hatching of fish larvae, and reduction of zooplankton competition to the fish larvae. However, this hypothesis was not verifiable with the plankton records.

The fluctuations of the Japanese sardine have been thought to be caused by fluctuations of larval recruitment. This sardine spawns in winter in Japanese coastal waters, and the centre of the spawning activity has moved slightly north and south in different periods. The present spawning ground in the northern part of Tosa Bay is thought to favour a longer residence time of the eggs and larvae in the bay when the westerly winds dominate during the spawning season and the path of the Kuroshio current meanders offshore south of Cape Maroto (Sugimoto *et al.* 1991). However, not all fisheries scientists concerned are convinced that this hypothesis is correct.

The possible effects of wind anomalies on the spawning success of the East Anglia herring was pointed out by Carruthers (1938). The fluctuations of the Georges Bank herring were also assumed to be largely influenced by the environment. In addition to fishing pressure, fluctuations of the herring stock on Georges Bank are believed to be caused by differential larval mortality (Lough *et al.* 1981). These authors believed that spawning activity may be enhanced by warmer autumn bottom temperatures. During the 1970s the centre of spawning activity shifted from eastern Georges Bank to western Georges Bank and the Nantucket Shoals. Currents were also different and the above authors believed that subtle changes in current patterns may cause extreme fluctuations in recruitment. Larval survival was higher during SW winds, which meant onshore transport. Gulf Stream rings and offshore wind driven transport were assumed to remove larvae from the Banks, which were then lost offshore and/or starved. However, no definite proof can be obtained for these hypotheses.

Recruitment of some demersal fish, such as cod, has also been believed to be related to winds. Martin & Kohler (1965) found that long-term trends and annual variations in recruitment of cod to the New England, Western Nova Scotia and Gulf of St Lawrence areas was negatively correlated with water temperatures during the first year of life. Temperature was taken as an indirect measure of the effects of wind and water transport of plankton, including eggs and larvae. Colder water implied greater water transport from east to west, which resulted in the appearance of strong year-classes on western grounds. On the other hand, Koslow *et al.* (1987) found that year-class success in northwest Atlantic cod stocks tends to be associated with large-scale meteorological patterns and offshore winds. These authors warn that it is possible to be misled on the basis of simple correlations into thinking that a particular variable underlies virtually any other in the region owing to widespread intercorrelations. They point out that it is unlikely that any of several plausible mechanisms is predominantly responsible for year-class success.

Eggs and larvae are subject to predation, less by their own kind, which usually leave the spawning grounds, than by other species that move into the spawning areas, often for the purpose of predation on eggs. The fecundities of most fish are high and can tolerate large losses of eggs and larvae. From about 2 000 000 eggs spawned by an average-sized cod, on average only two will survive to the age of 3 years. Although there will always be considerable mortality of eggs and early larvae, this mortality cannot alone regulate the remarkable semi-constancy of recruitment. Consequently there must be density-dependent regulating mechanisms in action in the few years between the larval stage and the age of recruitment. This density-dependent regulation mechanism is provided by predation. Fluctuations in predation are expected to be little influenced by climatic fluctuations.

Mass predation of larvae is known to occur at early larval and somewhat later stages, when there are mass developments of predatory large zooplankters, such as salps (doliolids), jellyfish (medusae) and ctenophores (comb jellies). At times it has been assumed that this type of predation might be a significant determinant of recruitment. However, quantitative data on the quantities and distributions of these predators are usually not available for reliable calculations of larval mortalities caused by them.

With increasing size, the larvae develop escape behaviour and subsequently shoaling behaviour sets in. With further increase in size the larvae and young fish become subject to hunting by other pelagic fish, but with still further increase in size, their escape velocity increases and the number of predators who are able to swallow the larger organisms decreases. Thus, there will be minimal total mortality before fishing mortality starts to take its toll and when spawning stress mortality sets in.

The predation on eggs and larvae has been summarized by Laevastu & Favorite (1988). It could be added that Ware (1986) found that most of the variation in Pacific herring recruitment off the Canadian coast can be linked to changes in predator (Pacific hake) abundance, zooplankton biomass and the particular pattern of moderate and strong ENSO (El Niño–southern oscillation) events that have occurred since the late 1930s. The predation by pelagic fish on the eggs of demersal fish may affect the recruitment of the latter. Daan (1976) showed that both herring and sprat may, under certain circumstances, concentrate on eggs of demersal fish for their food requirements. This points to the possibility of a direct causal relationship of the observed long-term antagonistic behaviour of the pelagic and demersal stocks in the North Sea. The predation by sea birds might also remove a considerable amount of fish larvae and small pelagic fish. Furness (1989) estimated that in the Shetland area the consumption by sea birds in 1981 removed 27% of the sandeel production.

To sum up, the spawner–recruitment relationship can be eliminated as a determinant for recruitment as too simplistic, except when the spawning biomass is very low. The relatively long spawning period and the space- and time-variable transport of eggs and larvae cause the first great uncertainty in the evaluation of recruitment. It is not certain that a major portion of the survival of larvae arises from the peak spawners or from major spawning grounds.

A largely unsolved problem is what happens to larvae which are swept offshore from coastal spawning areas. One hypothesis is that good recruitment will result from offshore larvae of some species, as the amount of predation is lower in offshore than in coastal waters.

Even more uncertain is the quantitative evaluation of predation during the year or year and a half after the larval stage. Here some quantitative data are needed about the density of the juveniles and the abundance of predators. Both are difficult, if not impossible, to obtain and can only be simulated in an ecosystem model. Thus the only course left for recruitment evaluation before full recruitment is to try to measure the relative abundance of pre-fishery juveniles, through young fish surveys.

Changes in distribution and migrations

Alterations of the distribution and of seasonal and life-cycle migration of fish might be caused by climatic changes, such as changing currents and advection of water types, concomitant with changes of temperature. However, there might be many other causes for changes in distribution, such as changes within the ecosystem, such as food availability and predation, or the effects of fishing. There are also changes in fish behaviour, e.g. selection of migration routes, for which we do not have good hypotheses or facts as to their causes. The distribution and migration changes are most pronounced in pelagic species, examples of which are given below.

The generalized distribution and migration of the Atlanto-Scandian herring prior to 1960 are shown in Fig. 5.10 (for history of stock sizes and catches see Chapter 7 section 2). The protected 1969 year class spawned at an early age in 1973, mostly on the coast of northern Norway (Bakken 1983). After spawning the herring did not migrate to the open sea, but remained along the coast during the feeding period in summer. In this way the life history and migration pattern of the stock was totally changed (Dragesund *et al.* 1980, from Bakken 1983), and no reasonable hypothesis for this change has been found.

The migration patterns of the Icelandic summer spawners also changed in the 1970s (Jakobsson 1982, quoted in Bakken 1983). In particular, the southward migration from the feeding area to the overwintering area off southern Iceland started earlier, and the stock remained during winter in dense shoals close to the shore in restricted localities on the south coast. The normal (1950 to 1965) migration pattern took the spent summer spawners northward off NE and NW Iceland during August and September. In late autumn (October to December) the stock was found off western Iceland and slowly migrated southward as the water temperature in this area decreased. In January or February the herring migrated to the area off Reykjanes and eastward along the south coast, and some summer spawners overwintered off southwestern Iceland.

Changes in distribution of migration and feeding grounds of North Sea herring also occurred in the 1980s. The North Sea herring spawning stock size and recruitment experienced considerable changes between 1960 and 1990 (Fig. 5.11). In the 1960s and earlier the herring larvae drifted from the

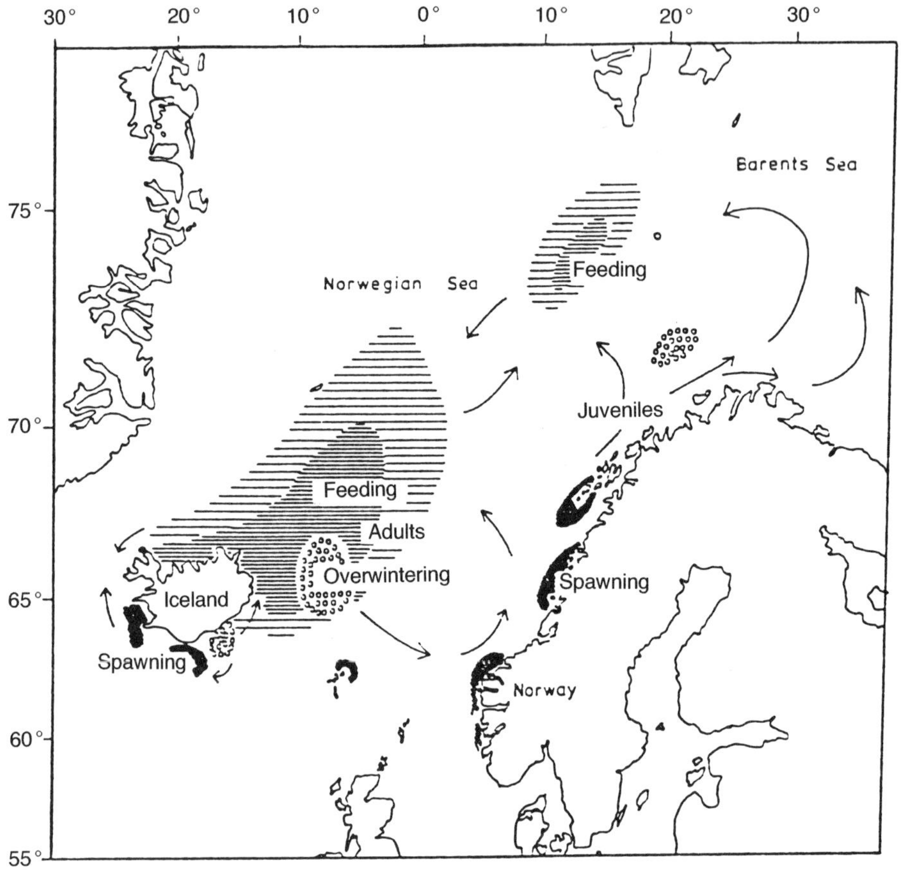

Fig. 5.10 Generalized illustration of the distribution and migrations of Atlanto-Scandian herring during a period of high stock level (Bakken 1983).

spawning areas in the west to the nursery areas in the east (Fig. 5.12) with the normal residual circulation in the North Sea. However, during the years of low recruitment in the 1970s the herring larvae did not reach the nursery areas in the eastern North Sea and in the Kattegat, whereas from 1980 on, with the increase of the stock, the appearance of older larvae in the eastern North Sea and Kattegat improved (Corten & van de Kamp 1991). It was speculated that the residual currents in the North Sea had changed during the 1970s. However, no proof of this change has been presented.

More remarkable is the shift of summer distribution of adult herring in the northern North Sea, described by Corten & van de Kamp (1991). During the 1950s, 1960s and 1970s the main feeding grounds for the herring were the waters off the east coast of Scotland, south of 37 °N and west of 2 °E (Fig. 5.13a). The distribution of the herring in the 1980s shifted north

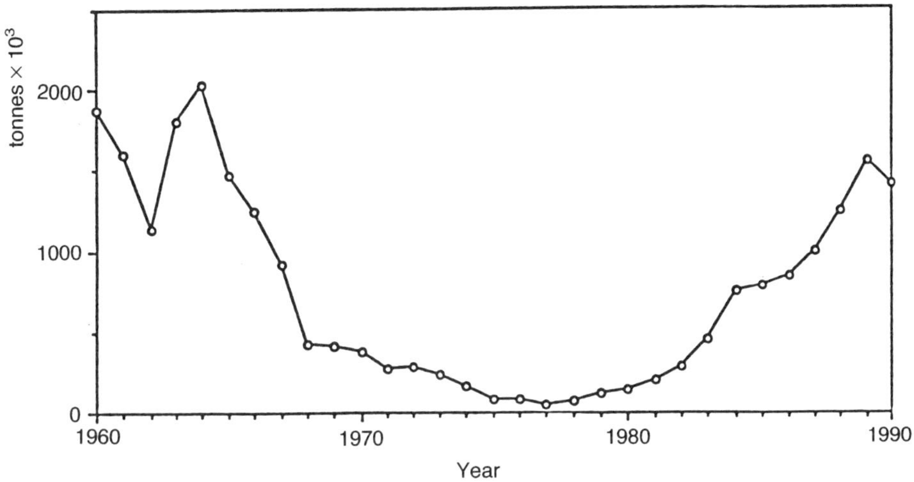

Fig. 5.11 Spawning stock size of North Sea herring. Data from earlier ICES working group reports (Corten & van de Kamp 1991).

and northeast to the edge of the continental shelf and to the Norwegian Trench (Fig. 5.13b). Similar dramatic changes in distribution of the western mackerel occurred in the same time period (Walsh & Martin 1986, reported by Corten & van de Kamp (1991)). The overwintering area of mackerel shifted from the southwest of England to the west coast of Ireland and to the north of Scotland, and the feeding area shifted from the waters around Shetland to the more eastern part of the North Sea and in some years northeast along the Norwegian coast.

It has been speculated that these changes were related to hydroclime changes, particularly to the increase in North Atlantic drift along the edge of the continental shelf. However, no hydrographic evidence of such change has been presented. On the other hand, Fraser (1961) found that herring in the northern North Sea do not appear in water of Atlantic origin which flows into the North Sea until this water mixes with North Sea water and rich zooplankton develops (see Section 2.6).

The study of long-term changes of the marine ecosystem in a given location is possible only if observations samples are collected routinely at relatively short time intervals over a long period of time. One of the very few such series of observations in the world is from Plymouth Laboratory of the Marine Biological Association of the United Kingdom. Weekly samples of plankton and of water for chemical analysis have been taken off Plymouth since 1924. Southward & Boalch (1986) reported on the observed long-term changes of the ecosystem off Plymouth in the English Channel in relation to fish populations and the following summary is derived from their report (see Fig. 5.14).

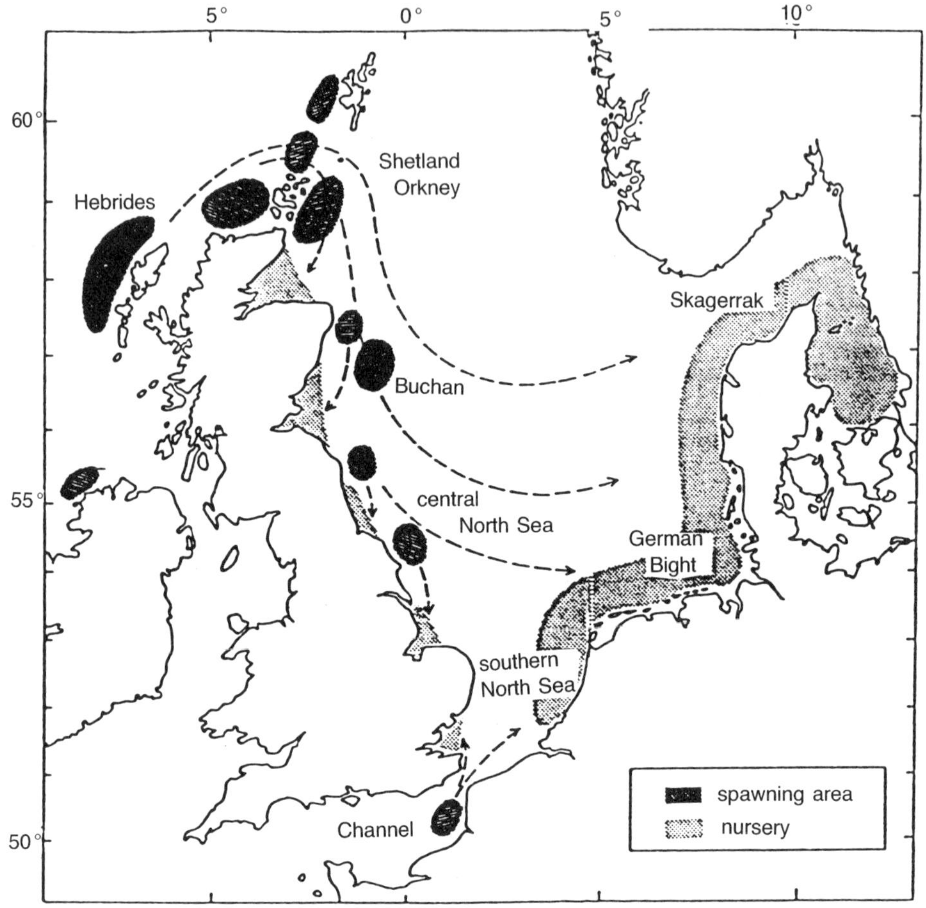

Fig. 5.12 Drift of herring larvae across the North Sea (Corten & van de Kamp 1991).

The abundance of zooplankton and clupeid postlarvae declined after 1930 by up to two orders of magnitude. This decline coincided with the collapse of the local fishery for herring. From 1936 onward the eggs of the sardine became abundant and the sardine also became a dominant pelagic species. Warm water species also became more prominent among demersal species in trawl catches. This changed ecosystem continued for about 30 years.

After 1965 the zooplankton increased again and by 1979 the planktonic ecosystem returned to the condition found before 1930, and the demersal fish of northern character became more common. The spring/summer spawning sardine decreased and virtually disappeared after 1975 (Fig. 5.14). From 1968 on mackerel replaced sardine as the dominant pelagic fish. After 1980 zooplankton and mackerel decreased again and the latter was replaced by sardine. Southward & Boalch linked these ecosystem changes only broadly

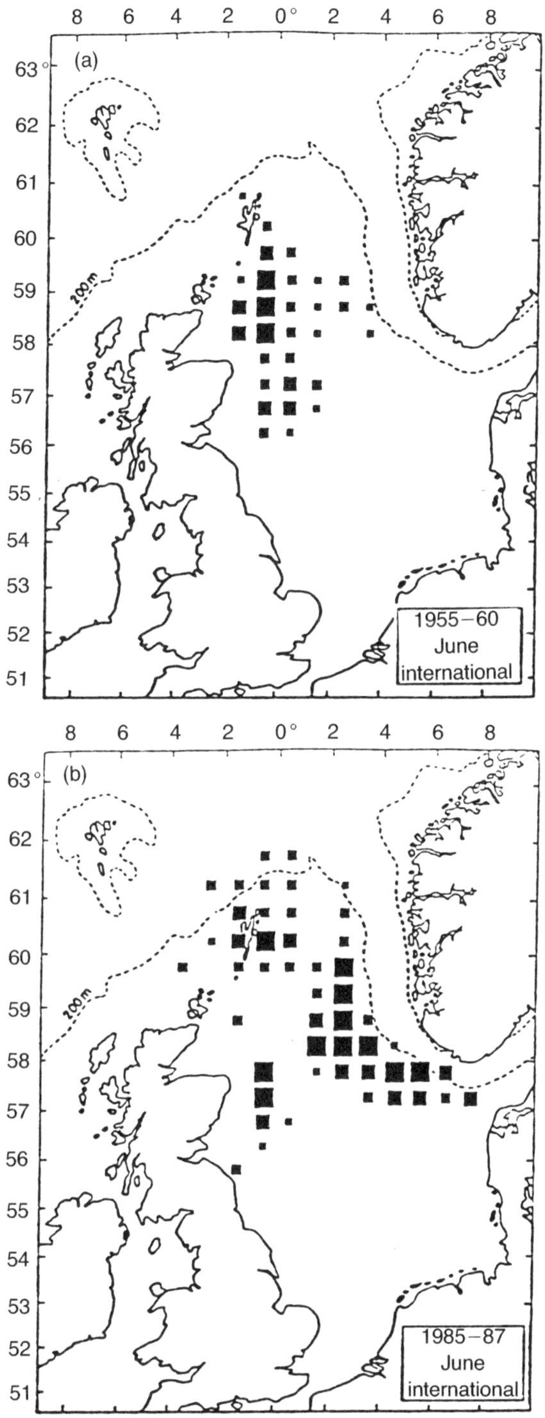

Fig. 5.13 Distribution of herring catches in June for two different periods. Black squares represent monthly catches of 250–500, 501–1000, 1001–2000, > 2000 t for the international fleet. Catches west of 4 °W are not included (Corten & van de Kamp 1991).

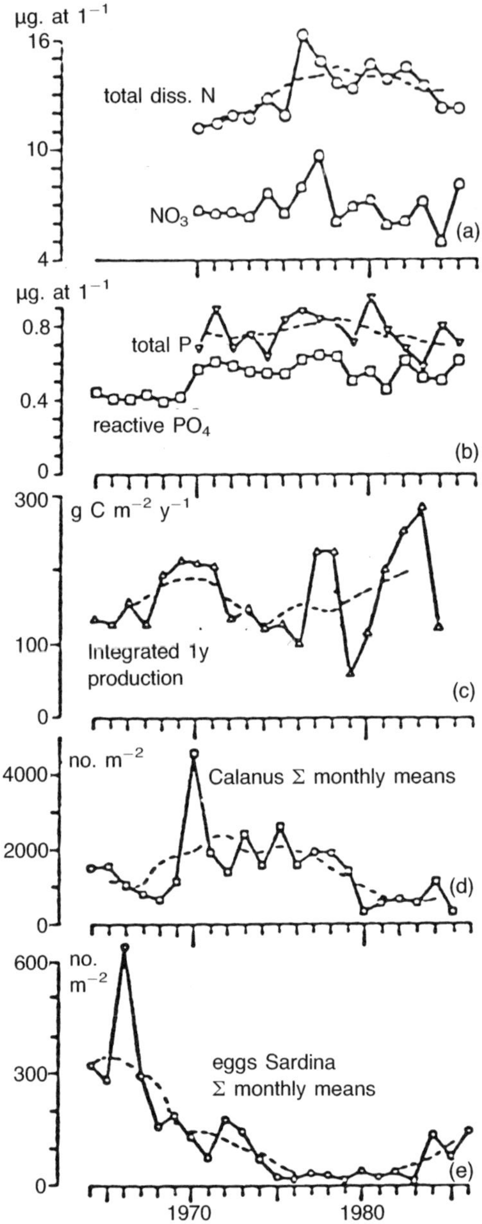

Fig. 5.14 Changes off Plymouth since 1964: (a) winter maximum of total dissolved nitrogen and inorganic nitrate at Station E1; (b) winter maximum of dissolved total phosphorus and reactive phosphate at E1; (c) integrated primary production at E1; as g carbon m^{-2} surface $year^{-1}$; (d) annual sum of monthly means of *Calanus helgolandicus* at station L5, as no. m^{-2} surface; (e) annual sums of monthly means of eggs of *Sardina pilchardus* at L5, — as no. m^{-2} surface. The broken lines are drawn through the 5-year running averages (Southward & Boalch 1986).

to fluctuations in sea temperature and climate. They found no evidence that the changes in ecosystem were related to basic productivity of the system or to influx of inorganic nutrients from outside the western Channel.

The French herring and sardine populations in the English Channel/Bay of Biscay coast are situated respectively at the southern and northern boundaries of their habitat. Thus climatic fluctuations might affect their distribution and catches. Binet & Orstom (1986) have summarized the fluctuations of these catches since the eighteenth century. They found that the southward shift of herring catches from Yarmouth to the Channel waters during the eighteenth century coincided with the climatic deterioration, and when the climate became warmer in the beginning of nineteenth century the Channel fishery declined. Good periods of Channel herring fishing occurred during climatic coolings and corresponded roughly to the good herring periods of the Bohuslän herring fishery.

The good and bad sardine fishery periods (Fig. 5.15) were probably related to both short- and long-term climatic variations. Binet & Orstom pointed out that although sardine fisheries in other parts of the world show somewhat similar trends, the Japanese, Californian and Spanish sardine catches peaked some 20 years before those of the French Atlantic fisheries.

The replacement of one species by another has been considered either as competitive replacement or as a shift of stock distribution under environmental influence. Although MacCall (1983) argued against competitive replacement, the two concepts are not mutually exclusive.

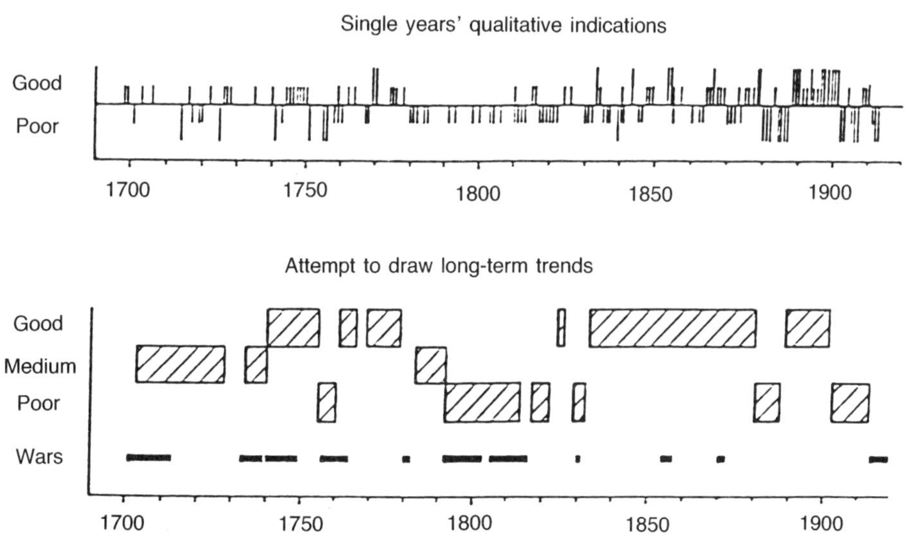

Fig. 5.15 Long-term trends of French sardine catches from the English Channel to the Bay of Biscay coast (Binet & Orstom 1986).

Geographic shifts of populations have been observed also in semi-demersal species. For example, Hempel (1978a) reported that the rise of the haddock stock in the North Sea in the 1960s was accompanied by a southward expansion into the southern North Sea. Many demersal species undertake seasonal migrations, which can be affected by environmental anomalies, especially in high latitude areas such as the Bering Sea (Favorite & Laevastu 1981). The variation of the extent of ice seems to have little direct effect on the seasonal migrations of demersal fish.

Semi-demersal fish can also alter vertical distribution under the influence of climatic fluctuations and thus change the availability to gear. MacPherson *et al.* (1992) found that abnormally warm conditions in the surface layers could induce cape hake to concentrate close to the sea bed, reducing the distance between individual fish. Such an effect would, of course, make the population more susceptible to bottom trawling.

Changes in growth and feeding

It can be postulated that climatic fluctuations might affect the abundance and species composition of the basic food of fish, i.e. zooplankton and benthos. Furthermore, it might be argued that the greater the ability of a fish species to consume a wide range of food organisms, the less it will be affected by the fluctuations in the abundance of any particular form of forage. If a species has catholic food habits, its predation is usually dependent on the density of the preferred prey present. This is the main density-dependent controlling mechanism of predation-controlled recruitment.

Icelandic fisheries scientists have studied the feeding conditions of herring during their extensive surveys. A detailed description of the results was given by Jakobsson (1979), a few essentials of which are pointed out here. Jakobsson found that considerable fluctuations of herring feeding conditions and habits occur during the normal and warm years. In 1960 the influx of warm Atlantic water off the north coast of Iceland was unusually strong, but zooplankton densities were low. As a result the herring were dispersed and chasing 0-group capelin at the surface, and this behaviour was affecting the herring catches. 1962 was considered a 'normal' year and high zooplankton concentrations were observed, especially in frontal zones and on coastal banks, which resulted in very high availability of the herring to purse-seines.

In the period 1964 to 1968 the Arctic waters of the East Icelandic current increased and transported drift ice north of Iceland. The primary production off the western and middle north coast dropped to only about a quarter of that in 1958 to 1964 and nearly complete collapse of zooplankton occurred. Radical changes of the migration of herring also occurred, no herring having been found in the zooplankton-poor water. Although primary production increased after 1970, zooplankton remained very low, indicating that the previous adverse conditions destroyed the *Calanus finmarchicus* population.

Scarnecchia (1984) reported that the marked change of hydroclime and associated decline of zooplankton and forage fishes north of Iceland from 1965 to 1970 caused also a decline in salmon yields from the Icelandic north coast rivers.

Growth rates of fishes are known to vary. The reasons can be varying food availability, a changing temperature regime, and/or some genetic changes in stock.

Bakken (1983) observed that the growth of the Atlanto-Scandian herring was considerably faster in the 1970s (8-year-olds 36.5 cm) than during the 1950s and 1960s (34 cm). The mean age of maturity was also reduced from 6 or 7 years to 4 years. A similar increase of growth rate in North Sea herring was observed earlier (Fig. 5.16, Hempel 1978a) which also resulted in sexual maturity being reached 1 year earlier, at 3 instead of 4. Hempel (1978a) furthermore analysed the increased growth rate of most species in the North Sea in the early 1970s during the high abundance of the species, but found no apparent reasons for this growth rate change and earlier maturation. Similar growth rate changes in the North Sea have been observed in other fishes, such as haddock. Tormosova (1983) believed that the warming of the North Atlantic from the 1930s to the 1950s caused an increase in growth rate and earlier maturation of haddock. This belief cannot, however, be verified to a satisfactory degree. On the other hand, Ware (1986) believed that the

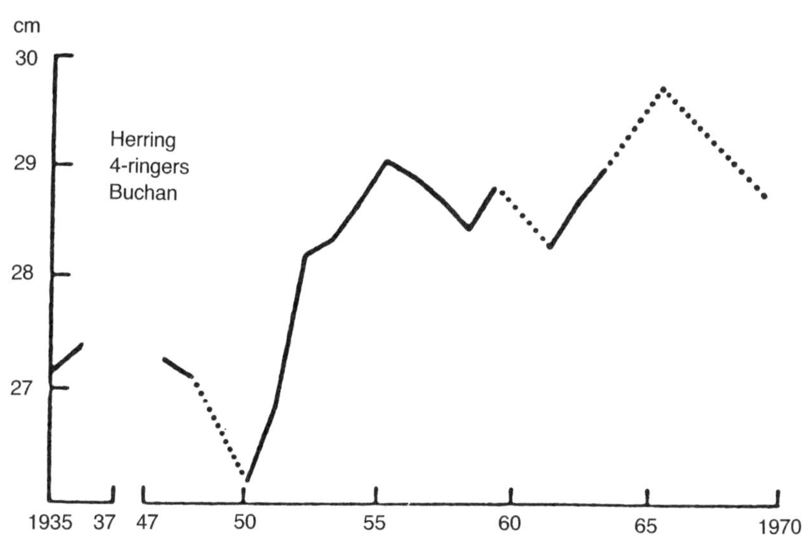

Fig. 5.16 Increase in length of 4-year-old Buchan herring (Hempel 1978a).

increased growth rate of Pacific herring was caused by variation in zooplankton biomass and by inter-annual variation of herring year-class size.

As long as we cannot test the many hypotheses on the causes of observed growth rates of fish, we cannot associate quantitatively any growth rate alteration directly with hydroclime changes.

5.3 Effects of climatic change on demersal stocks

Climatic fluctuations affect mainly the surface layers of the sea. One might think, therefore, that demersal fish are less affected by climatic fluctuations than pelagic fish. However, many demersal fish have pelagic larvae which are subjected to environmental conditions in the surface layers.

One of the classical examples of the effect of climatic fluctuation on fish and fisheries has been the west Greenland cod fishery. A summary of the changes of this fishery is given below based on Buch & Hansen (1986).

A rich cod fishery developed west of Greenland from the late 1920s onward, and during the 1950s and 1960s the catch was more than 300 000 t (Fig. 5.17). The catches declined rapidly in the late 1960s and present catches are less than 20 000 t. The occurrence of an abundant cod stock west of Greenland coincided with a warmer period in this area (see Fig. 5.18). When the cooling started around 1970 catches started to decrease nearly instantaneously, and when the second strong cooling occurred in 1982–1983 catches of cod declined to almost nil. Buch & Hansen (1986) found that

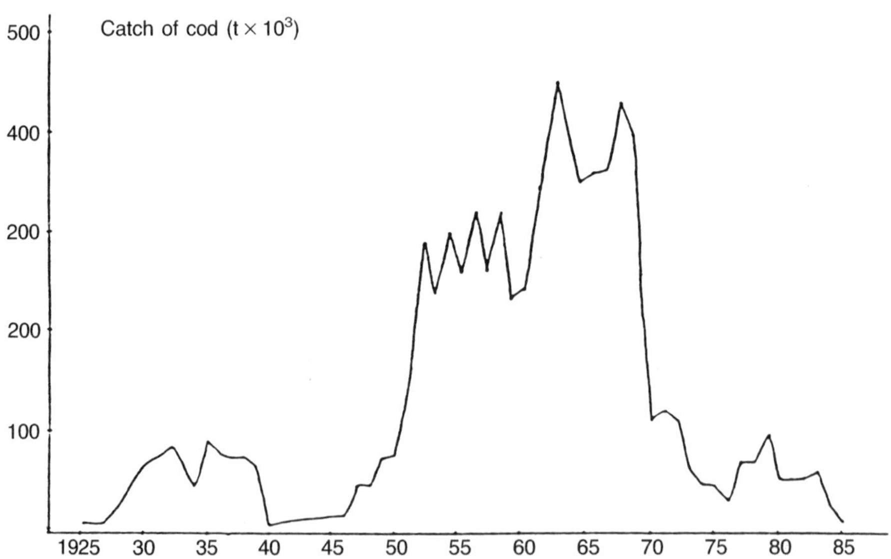

Fig. 5.17 Catches of cod off West Greenland, 1925–85 (Buch & Hansen 1986).

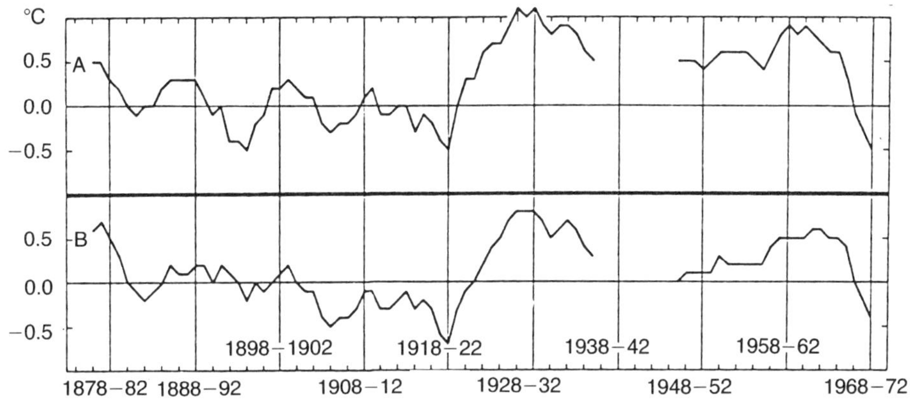

Fig. 5.18 Five-year overlapping means of annual sea surface temperature anomalies in areas A off the West coast of Greenland and B off the south coast of Greenland (Smed 1980).

intensive fishing reduced the cod stock but that the actual collapse of the fishery was caused by the failure of recruitment and possibly the supply of larvae and juveniles from Icelandic spawning grounds. Local climatic deterioration might also have caused the displacement of the stock southward.

Spawning of cod occurs in a relatively continuous belt south and west of Iceland and also southeast and southwest of Greenland. It has been found that recruitment to the west Greenland cod stock is highly dependent on a larvae/young fish drift from Iceland and less dependent on spawning in Greenland waters (Buch & Hansen 1986). This led to the hypothesis that the collapse of the cod stock of West Greenland is coupled to variations in the number of young fish recruited from Iceland. This transport is effected by the Irminger Current (Fig. 5.19). The Irminger current is mainly wind driven and its variation is thus dependent on climatic fluctuations in the surface wind regime. The fluctuations of the Greenland cod stocks is one of the few plausible examples of the effects of climatic changes on fish stocks.

There is also an emigration of old cod from West Greenland to Iceland, ascertained by tagging experiments. Furthermore, a southward placement of the West Greenland stock has been observed as the cod in West Greenland is at the northern boundary of its distribution. Its growth rate is also changing according to temperature conditions.

In the 1960s there was a considerable increase of gadoids in the North Sea (Fig. 5.20). Andersen & Ursin (1977) believed that this was caused by a decline of mackerel and herring stocks which was mainly due to heavy fishing. As a result the predation on gadoid larvae was decreased, allowing improved recruitment. In addition, more food might have been available to juvenile gadoids. Cushing (1980), however, raised the possibility that the gadoid upsurge might have a climatic origin, but there are no hydroclime

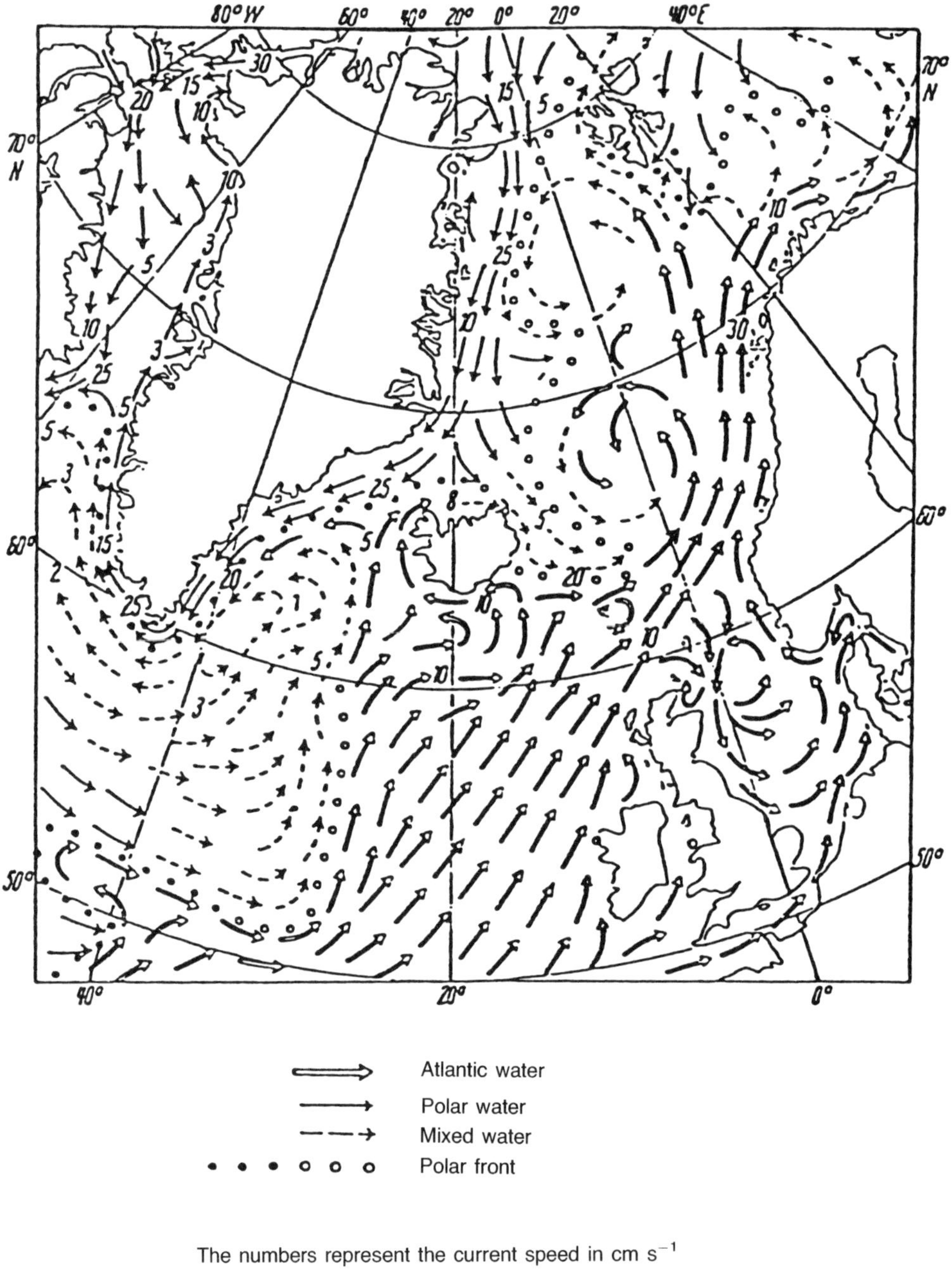

Fig. 5.19 Surface currents in the northern part of the Atlantic Ocean (Buch & Hansen 1986).

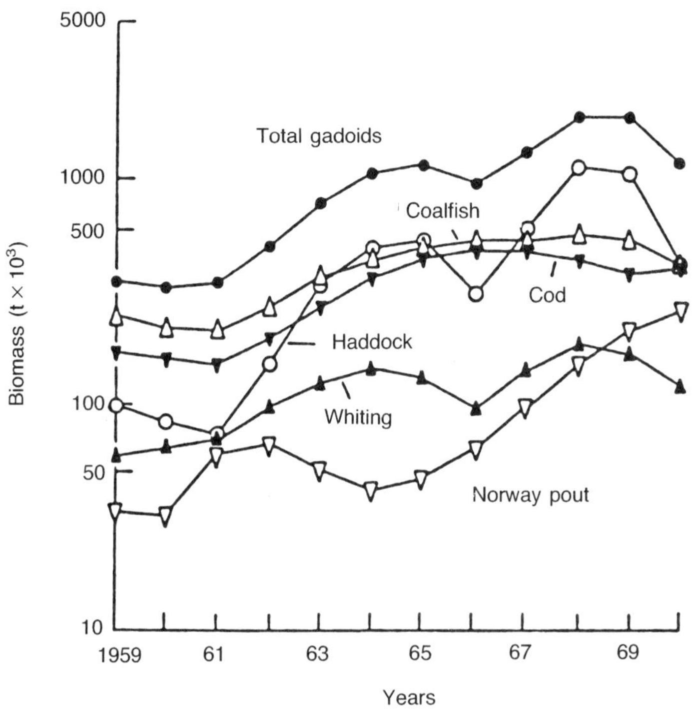

Fig. 5.20 The gadoid outburst in the North Sea, biomass in thousands of tonnes, logarithmic scale (Cushing 1980).

data to support this. Lee (1978) raised three possible reasons for the increase of gadoid stocks in the North Sea. First, there has been an increase in nutrients (e.g. phosphates) in the North Sea, which might have resulted in higher plankton production and consequent improvement in the year-class strengths of various species of roundfish. The second possibility is that the marked decrease in pelagic stocks has been compensated by an increase in demersal stocks, and third is that a cooling of climatic conditions over the northern and central North Sea has led to improved year-classes of demersal fish. However, the quantitative contribution of any of the possible factors to the changes of gadoid stocks cannot be ascertained.

The changes in demersal fish stocks on Georges Bank have been described by Overholtz & Tyler (1985), based on spring and autumn surveys. The decline of major commercial species in late 1960s and early 1970s is ascribed to heavy fishing pressure. The increases or decreases in species within the ecosystem were assumed to be controlled mainly by processes within this system, mainly predation and competition for food. No direct climatic effect could be determined as the cause of the changes.

A different example of variation in demersal fish in response to climatic fluctuations can be brought from subtropical upwelling regions. During near-cessation of upwelling (e.g. during El Niño 1982–3), demersal fish benefited from improved oxygen content and feeding conditions at the sea bed and extended their range of distribution toward shallower water and to the upwelling area and increased in abundance (Arntz 1986).

5.4 Marine mammals and birds and fisheries

Before man intensified fishing and drastically reduced marine mammal herds, the mammals were the top predators on fish. At the present time herds of larger mammals are still found in the Antarctic, and during the summer the fish-rich Bering Sea contains the greatest numbers of marine mammals and birds in the northern hemisphere. Laevastu & Marasco (1992, in press) conducted a quantitative study on trophic interactions of marine mammals and birds with fishery resources in the eastern Bering Sea. Their essential results are given below:

Two million tonnes of fish are eaten annually by marine mammals and birds in the eastern Bering Sea. The same amount of fish is taken by the commercial fishery. The annual removal of fish by marine mammals and birds is 4.4% of the mean annual standing stock of all fish. The consumption of fish by birds is less than 10% of the consumption by mammals. About half of the fish eaten by mammals is of commercial interest. One per cent (or 130 000 t) of commercially valuable fish eaten by mammals is of commercial size, the rest being pre-fishery juveniles. If 50% of these pre-fishery juveniles survive and grow to exploitable size, 1 110 000 t of exploitable stock, or 1 230 000 t including consumed exploitable size fish, would add to the presently available exploitable stock. This amount is 6 to 12% of the exploitable stock present.

Mammal and bird predation on fish is the equivalent of 2 160 000 t of fish, i.e. these fish would eat the same quantity of fish as mammals and birds eat now. This additional (equivalent) fish biomass is only 4.8% of the present fish biomass in the eastern Bering Sea (44 700 000 t). There is little, if any, competition between mammals and birds and commercial fishing, as mammals and birds consume mainly small and noncommercial species. One notable exception is returning salmon, of which about 30 000 tonnes are taken by mammals, mainly by Beluga whales, Steller's sea lions and larger fur seals.

The total fish-carrying capacity of the eastern Bering Sea is determined by the production and availability of the basic food resources, zooplankton, benthos, and fish. Most mammals and birds utilize the same food resource as the fish ecosystem. Thus the comparative effect of marine mammals on the basic food resources would be similar to the increase of total fish biomass with equivalent food consumption (2 160 000 t). The consumption of zooplankton by mammals and birds is only 1.6% of the consumption of zooplankton by the fish ecosystem; the corresponding percentage for benthos is 5.5.

The removal of fish by the fishery does not affect the food resources of marine mammals, as fishing removes the larger, more piscivorous fish whose diet is similar to that of the major mammal component. The recruitment to the fishery is in most species in the Bering Sea determined by predation on larvae and juveniles by the fish ecosystem itself and the corresponding predation by mammals and birds is insignificant.

The interactions between marine mammals and fishing have recently been summarized by Alverson (1991). These interactions have little effect on marine mammals in terms of kills. However, marine mammals are a nuisance to fisheries. The catch is taken by mammals from long-lines during hauling. Some mammals get caught in trawls during inhauling, and to release them costs valuable ship time. Therefore it is desirable to develop methods to keep marine mammals at a distance from fishing vessels.

In the Baltic Sea there has been a considerable increase in primary production (see Chapter 6 section 2) and a corresponding increase mainly in pelagic fish production. At the same time the number of Baltic seals has decreased for various reasons, including diseases and the activities of man in the coastal breeding and hauling areas. Elmgren (1989) calculated that in 1900 marine mammals needed about 5% of the primary production of the Baltic to sustain their food chains, whereas the fishery required only about 1%. Today the few remaining seals require less than 0.1% while the fishery needs about 10% of primary production for the same purpose.

A different effect of marine mammals has recently occurred in the Barents Sea (see Chapter 7 Section 2). The total fish ecosystem decreased there drastically in late 1980s. As a result seals did not find sufficient food on traditional feeding grounds in the Barents Sea and invaded northern. Norwegian coastal areas, where they were of considerable nuisance to local fisheries.

Crawford *et al.* (1987) reported a very large recent increase in the population of Cape fur seals, which was considered a recovery after decreased exploitation. It is assumed that considerable amounts of Cape hake are consumed by these seals.

In 1970s and 1980s a considerable decrease in numbers of Steller's sea lions has been observed in the Bering Sea and Gulf of Alaska (Alverson 1991). Harvesting by Eskimos and incidental catches by trawls cannot account for this decrease. Instead Alverson has convincingly shown that this decrease might be greatly effected via changes in composition of the food of mammals caused by changes in dominance of forage species. The stocks of herring and capelin have drastically decreased, whereas pollock and cod have increased. Herring and capelin, which have a fat content more than twice that of the pollock, used to constitute the main food resource for sea lions. However, meagre pollock might cause fat deficiency and increase the mortality of Steller's sea lion.

No weather or climatic effects can be found to be associated with the fluctuation of marine mammal stocks. The recent migrations of seals from the Barents Sea to the Norwegian coast, associated with the change of the Barents Sea fish ecosystem, has little if any connection with local climatic and hydroclime fluctuations.

Chapter 6
Pollution and Fisheries

There are two main man-made effects on fish stocks, namely fishing and pollution. The effect of fishing on fluctuations of stocks has recently been reviewed by Laevastu & Favorite (1988) and an additional brief summary is found in Chapter 5 section 1. This chapter reviews briefly the effect of pollution on fish stocks and attempts to separate this effect from the effects of climatic fluctuations.

Pollutants can be harmful (e.g. causing diseases in fish and/or man) or useful (e.g. eutrophication of semi-closed seas). It must be borne in mind that sea water contains nearly all the elements in solution, and that these elements can accumulate in marine organisms in varying degrees. This basic knowledge has sometimes been overlooked, as in the mercury scare in the USA a few decades ago. This section is not concerned with chemical pollution and the pathways of elements through the marine food chain, but reviews the effects of pollution at large on the fish ecosystem.

Another fishery-specific pollutant has received renewed attention lately, the lost and discarded nets and other gear (Anon 1991). There is a concern that some of the gear continues to catch fish; however, the quantities caught are small. This problem is not discussed further in this book because it is only very marginally related to weather.

6.1 Pollution in coastal waters and in semi-closed seas and its effects on fisheries

Most man-made pollution reaches the sea from the coast and the rivers. The pollutants are diluted by mixing processes in the sea, and are removed by chemical decay, biological uptake and sedimentation processes. Coastal pollution and its effects are local, highly varied and most pronounced in estuaries. One of the greatest effects of local marine pollution has been on marine recreation. The water quality on the recreation beaches is usually ascertained by the amount of *Escherichia coli* and enterococcus densities and by considering their effects on man through gastro-intestinal disease (Cabelli *et al.* 1976).

The most direct effect of coastal pollution appears in the frequency of diseases infectious to man which are transmitted by mollusks and crustaceans.

However, there are also naturally occurring diseases and parasites in these organisms, which are infectious and can occur periodically. Some of the disease organisms, including viruses, may be latent in shellfish and may be activated by the presence of certain chemicals (Rosenfield 1976). Dethlefsen & Tiews (1985) pointed out that it might be doubtful whether a high prevalence of certain diseases in some species in the North Sea can be pollution-induced, as they are found in large areas inshore as well as offshore. A few of the recorded diseases can cause rare mass mortalities of organisms in limited areas. A mixture of pollutants exists in some coastal areas and it is impossible to attribute to any single contaminant a potential for impact on the marine population at large.

Estuaries receive a variety of contaminants from rivers which might affect the benthos and fish fauna in estuaries and in nearby coastal areas. Tiews (1983) reported a 28-year study of the fish fauna in the German Wadden Sea, showing drastic changes in the fish bycatch of shrimp fisheries. The traditional Wadden Sea species decreased and intrusion of non-traditional species increased. He could not explain the changes in species composition as possibly caused by hydroclime and/or climatological change, or as caused by fishing, but assumed that coastal pollution might have played a role in them.

Lee (1978) pointed out two other possible local effects of man's activity on coastal fish populations; land reclamation, e.g. in Netherlands Waddensea, which might deprive sole, plaice and cod of important nursery areas; and marine sand and gravel extraction, which might alter benthos, destroy bottom attached fish eggs and increase siltation so as to suffocate shellfish. The extent of these possible effects on fish stocks has not been estimated quantitatively. The areas affected by these activities of man are relatively small, and the total effects on the stocks can be expected to be small.

Trawling with beam and otter trawls might also affect benthos and the demersal fish ecosystem. A study of the ecological consequences of dredging and bottom trawling in the Danish Limfjord by Riemann & Hoffmann (1991) found that this activity considerably increased the amount of suspended matter in the water, and increased the internal nutrient load, oxygen consumption and possibly phytoplankton primary production. Some of these effects are similar to those which occur with heavy wind and wave action in shallow water. A summary review of the effects of trawling on benthos and fish is given by Laevastu & Favorite (1988), who concluded that these effects on fish stocks are small indeed.

The effects of marine pollution are local, confined mostly to estuaries and to the vicinity of marine outfalls. The semi-quantitative separation of their small and local effects on local marine fish ecosystems from the possible effects of fishing and climatic changes is possible, but these effects on a regional basis are usually small and negligible. However, many claims of the effect of marine pollution, often unsubstantiated, appear in the news media.

6.2 Eutrophication and fish production

Eutrophication is a term borrowed from limnology, and its modern application to the sea is, strictly speaking, incorrect. The classification of lakes recognizes eutrophic (highly productive), oligotrophic (poorly productive), and dystrophic (acid, humic) lakes. It has been assumed that eutrophication might be taken to mean the making of a lake (or the sea) more productive by adding limiting nutrients such as phosphates and nitrates with e.g. domestic or agricultural waste waters. However, eutrophication causes an eutrophic lake to be more eutrophic but does not always make an oligotrophic lake eutrophic. Thus the correct term should be auxotrophication. However, as the word eutrophication is widely in use, we will use it in this book as well.

Eutrophication of marine waters is noticeable in some estuaries and in semi-closed seas with limited flushing and with very small tides, such as the Baltic and Black Seas and the Mediterranean Sea. In the North Sea, with heavy tides and relatively good flushing, eutrophication is minimal. It was assumed in the 1960s that the increase of some fish stocks and catches from the North Sea might be caused by eutrophication. Later investigations could not verify this assumption. Hempel (1978b) reported that basic food (phytoplankton) production in the North Sea is normally not a major limiting factor in the productivity of the North Sea fish stocks and possible eutrophication effects would be minimal. After extensive Dutch plankton monitoring studies Colijn (1991) reported that eutrophication effects along Dutch coast generally extend only to a narrow stretch along the coast or area of limited size. In most areas these effects are hardly distinguishable from natural variation.

Some increase in benthic invertebrate biomass has, however, been observed in the Skagerrak–Kattegat (Josefson 1990), which has been assumed to be caused by increased land fresh water run-off from western Sweden and possibly from the Danish inner waters which might have influenced the input of nutrients to the sea. The effect of Baltic water flowing into these seas can also not be ruled out.

Eutrophication is a widely-reported phenomenon in the Mediterranean coastal zones (Caddy & Griffiths 1990), especially in Gulf of Lions and in the Aegean Sea. Most of the pollution effects in the Mediterranean have been noticed close to the sources of pollution, sewage run-offs and polluted rivers. Contrary to general eutrophication effects, these local pollutions affect tourism, marine recreation and coastal fisheries. The most obvious effects are algal blooms, including toxic dinoflagellates, and resulting anoxic conditions near the bottom, and associated fish kills and possible viral infections from consumption of contaminated shellfish.

In some areas in the Mediterranean, Adriatic and Aegean Seas the eutrophication extends offshore, increasing primary production and associated production of pelagic fish, such as anchovies.

The Black Sea receives untreated sewage and industrial pollution from the countries surrounding it. A further source of eutrophication in the Black Sea is the mixing of oxygen-void deep water containing hydrogen sulphide and high amounts of nutrient salts into the surface layers. Furthermore, the water exchange (flushing) in the Black Sea is dependent on run-off, which has been changed by man through river diversion into the Caspian Sea.

The Baltic Sea receives treated sewage effluent as well as nutrients from agricultural run-offs from rivers. The accumulation of these nutrients in the surface layers depends, besides sedimentation processes, on the water exchange with the Skagerrak–Kattegat, which is dependent on climatic conditions as well as on river run-off. The latter is in turn dependent on climatic fluctuations as well as on man's changing use of the water for irrigation and domestic use and on land damage and forest practices.

The changing river run-offs from three rivers running into the Baltic are shown in Fig. 6.1. (Mälkki & Tamsalu 1985). Comparison of river run-off with temperature fluctuations in Sweden and in the Baltic (Fig. 6.2), which is another indication of some climatic fluctuation, shows little, if any, relationship. Firstly, the river run-offs are influenced by man's activity. Secondly, the climatic changes in precipitation and consequent river run-off and climatic temperature changes are not necessarily simply linked, permitting a direct graphical correlation to be shown. A common trend can, however, be recognized between 1925 and 1950, a gradual decrease of run-off and an increase of air and water temperature. Furthermore, the supply of nutrients by rivers to the Baltic Sea is not necessarily a function of river flow, as the concentration of the nutrients in river water can vary.

A summary of the sources of nutrients to the Baltic Sea and their accumulation has been given by Elmgren (1984), showing that the main sources of both phosphorus and nitrogen compounds are rivers, whereas the greatest removal of phosphorus occurs by sedimentation, and of nitrogen by water exchange through the Danish Sounds and denitrification from the sediments. Whether man has influenced the fluctuation of the amount of anoxic bottom water in the deeper parts of the Baltic is disputable. A steady state between nutrient source and sinking might have developed in the Baltic Sea since 1980 (Fig. 6.3). However, the stagnation of bottom water has persisted since 1977.

Elmgren (1989) found that in twentieth century the eutrophication of the Baltic has increased pelagic primary production by 30–70%. Zooplankton production has increased less, but fish catches have increased tenfold, which is due only partly to increased eutrophication but also to increased effort and decreased number of marine mammals.

The commercial catches from the Baltic doubled from 1965 (450 000 t) to 1980 (900 000 t), remaining constant thereafter (Nehring 1991). The growth rates of sprat and herring have increased, showing improved feeding con-

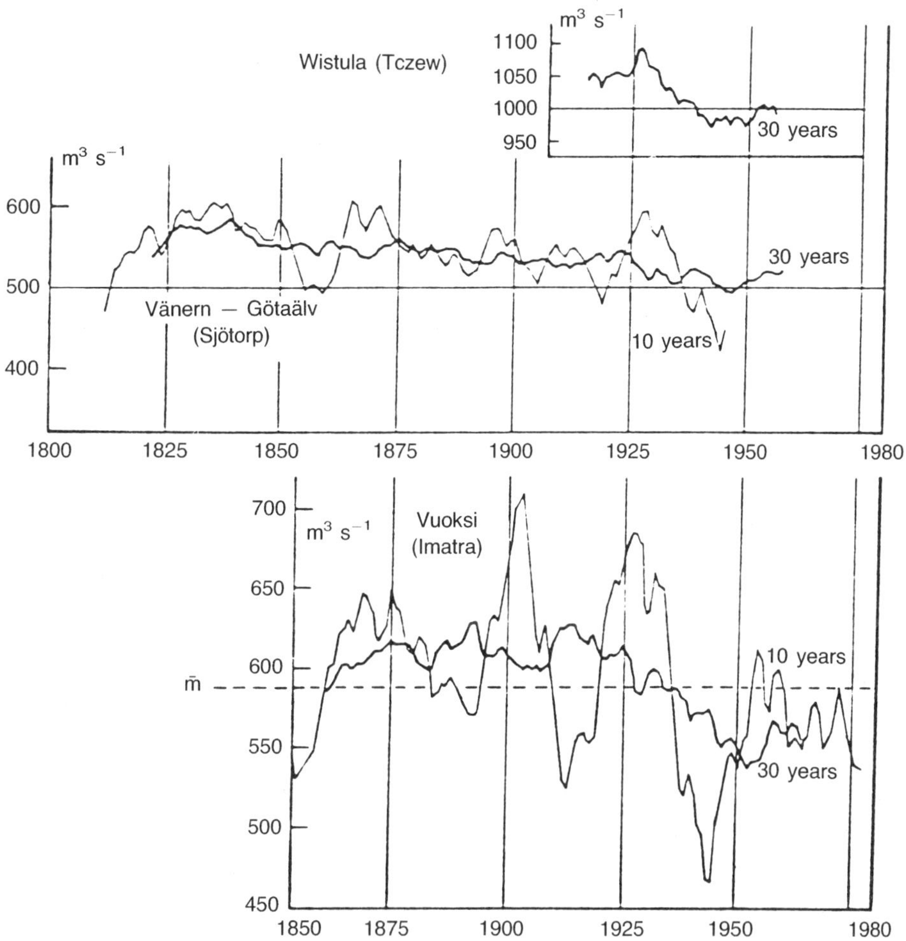

Fig. 6.1 Long-term fluctuations of the run-off of some rivers, 10 years and 30 years running means (Mälkki & Tamsalu 1985).

ditions. With the increase of cod stock in the southern Baltic, herring and sprat decreased in the 1980s, presumably owing to predation.

Cycles of about seven years in the atmospheric circulation, river discharges (including nutrient supply trends), and ice cover can be recognized in the Baltic. According to the interaction of several processes, Kalejs & Ojaveer (1989) classified major periods of environmental variations in the Baltic by their length (major *c.* 23 years, minor 7 and 14 years) and by various events;

(1) in periods of low fresh water inflow deep water influx from the Kattegat intensifies;

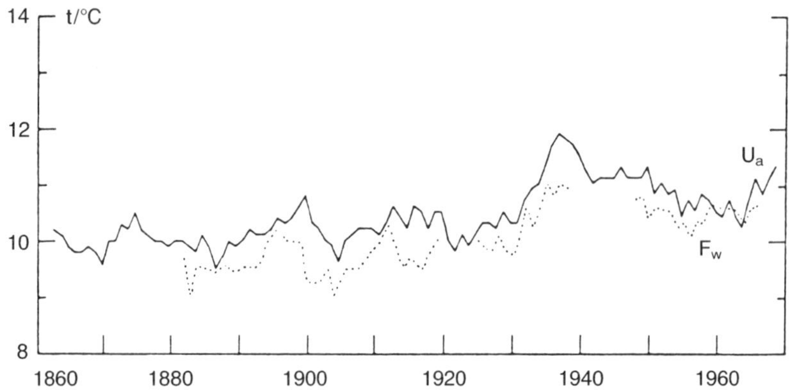

Fig. 6.2 Air temperatures at Uppsala (U_a) and water temperatures at Finngrundet (F_w) during summer months (Mälkki & Tamsalu 1985).

(2) in periods of intense fresh water inflow the Kattegat influxes decrease;
(3) in periods of mild winters the total heat content increases; and
(4) the opposite happens in periods of severe winters.

The pelagic fish respond to these changes in a relatively complex manner (Fig. 6.4) and causal relationships are difficult to detect.

The Baltic Sea is one of the semi-closed seas which is most sensitive to climatic changes as well as to man's actions. Furthermore, long time series of good data are available from this area. The available analyses show climatic and corresponding hydroclime fluctuations on all time scales. However, man, through eutrophication and intensified fishing has had the greatest effect on the fish ecosystem, and this effect has been positive indeed.

6.3 Effects of oil development and oil pollution on fish, shellfish and fisheries

Much attention has been given in the last two decades to potential effects of oil pollution on marine fisheries (e.g. Wolfe 1985). The pollution is thought to arise from accidents during transport (tanker accidents, Fig. 6.5), spills during loading or unloading, and well blow-outs. Voluminous literature exists on accidents and their effects. Mostly the effects are local and complete recovery from an acute local oil pollution of pelagic environment is rapid; the recovery of the benthic environment takes 3 to 10 years (Gray 1982).

Dethlefsen & Tiews (1985) considered the danger of low-level chronic oil pollution in the North Sea to be relatively low. Similarly Johnston (1977) concluded that loss of fish production or its approximate cash equivalent due to oil development in the North Sea, and possible accidents resulting from

Map of stations

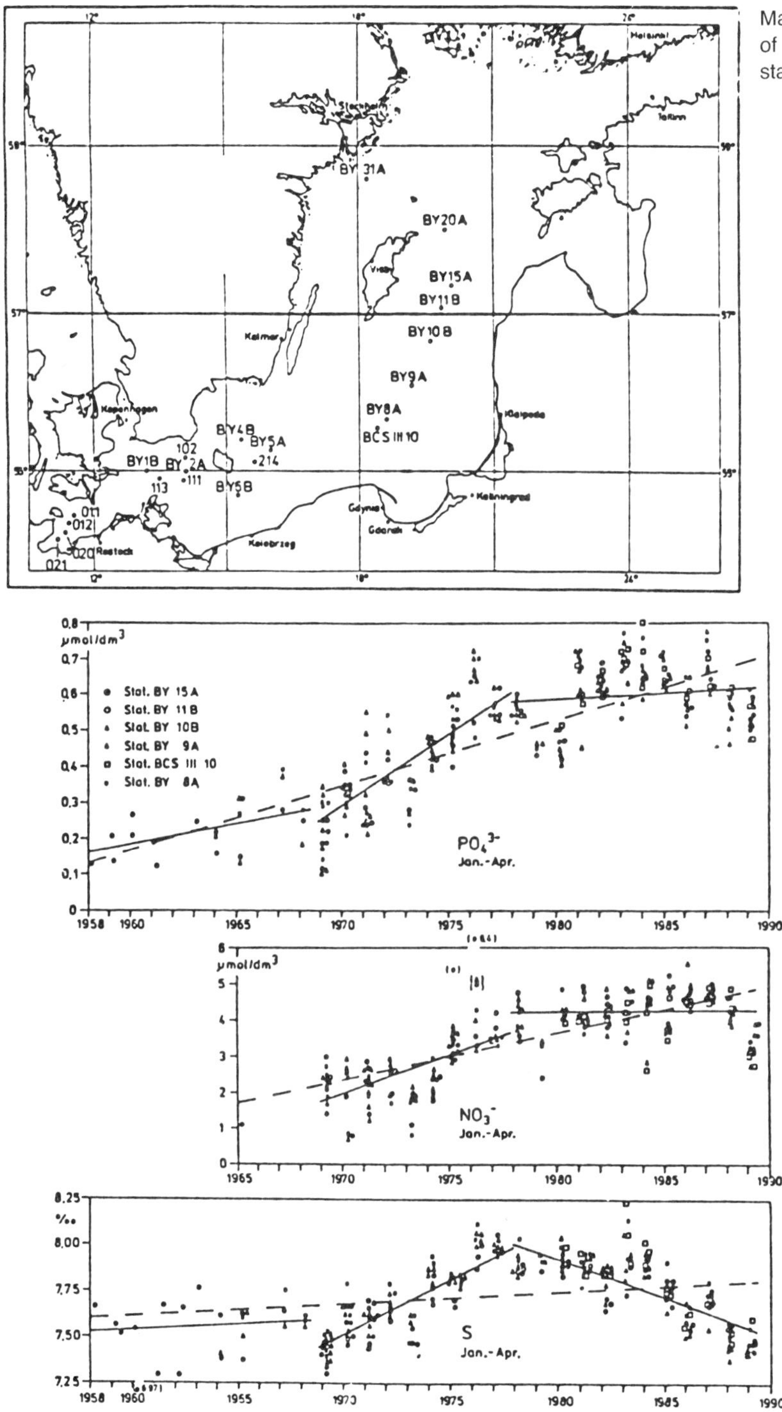

Fig. 6.3 Long-term trends of nutrients and salinity in the surface layer in winter at stations in the central Baltic Sea (Nehring 1991).

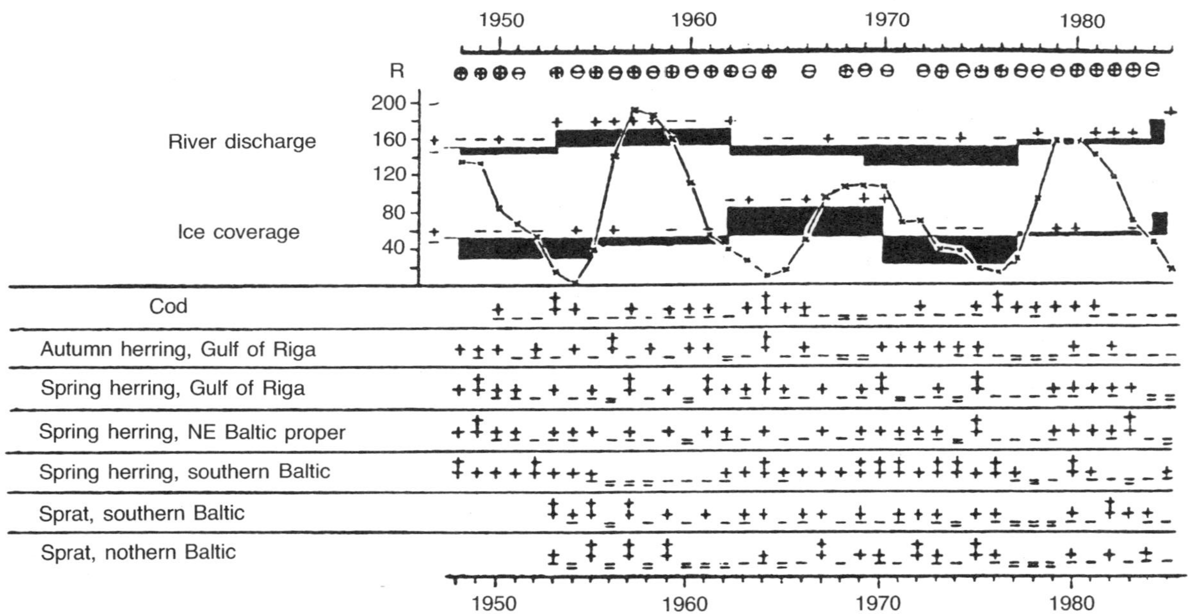

Fig. 6.4 The main regime-forming factors and year-class abundance of commercial fish populations in the Baltic in 1948–83. Dominating (more than 45% by strength) intensity of winds of western (WNW to WSW (+)) and eastern (ENE to ESE (−)) bearing in January–March. River discharge and ice coverage – in addition to the periods and their deviations from the long-term average, the years with higher (+) and lower (−) corresponding feature values are indicated. Sun activity (Wolf numbers x−−x) – R. The year-class abundances according to a five-unit scale: abundant (++), above average (+), average (+−), below average (−), poor (=) (Kalejs & Ojaveer 1989).

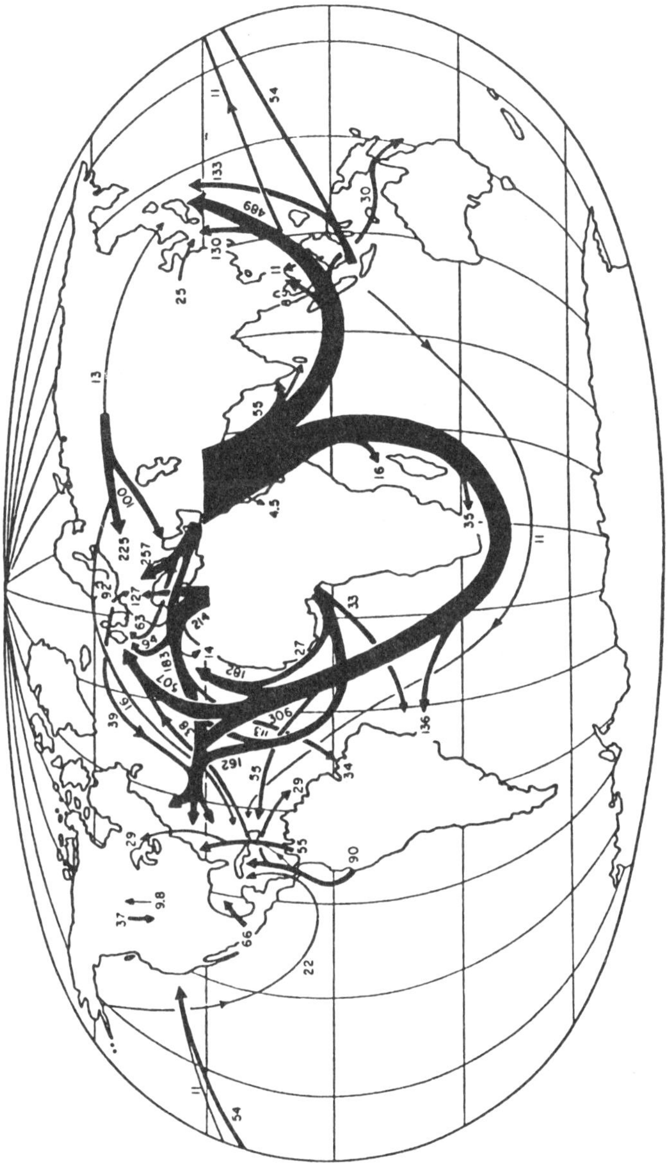

Fig. 6.5 Crude oil transport on the world oceans for 1979, in thousands of tonnes per day. The arrows indicate origins and destinations, but not necessarily specific routes (Wolfe 1985).

it, were small even for a catastrophic oil spill. There might, however, be some obstruction to fishing preventing contaminated fish from reaching markets. Only minor effects of oil developments on benthos near the Ekofisk and Eldfisk platforms in an area with a radius not exceeding 2 to 3 km have been found (Gray *et al.* 1990).

Laevastu & Marasco (1985) conducted an extensive evaluation and literature survey of oil development on the commercial fisheries in the eastern Bering Sea. The following conclusions are extracted from this study.

Laboratory experiments into the effect of oil on marine biota are well covered in the literature. Field application of this knowledge is, however, on many grounds difficult and questionable. The lethal and sub-lethal effects of a possible major oil spill on eggs, larvae and fish on the principal fishing areas, such as eastern Bering Sea, were found very small in relation to resource and limited in time and area. The greatest effect of a spill might be on beaches if these are used for holiday-making. The longest lasting effects might be on benthos from weathered oil on the bottom. This effect might last for a few months and cause tainting of fish which feed on the contaminated benthos. Areal coverage of benthos contamination is, however, small and natural purification rapid, at most a few weeks.

An economic effect to consider in the rare event of an accidental oil spill might be the loss of fishing area for a few months to prevent the fishery from catching tainted fish. The greatest damage to fisheries is done by environmentalists and sensation-seeking news media, which might affect marketing of local fresh fish.

Chapter 7
Major Historical Changes in Fisheries and Expected Future Trends

7.1 Global and large-scale changes

In the past decade the notion of global warming has received much attention. There is no doubt that owing to increase of 'greenhouse gases', mainly carbon dioxide and methane, in the atmosphere, caused by man's increased use of fossil fuels and by deforestation, a slight increase of global temperature is occurring. As there is an exchange of these gases between the atmosphere and the ocean, it can be assumed and computed that an equilibrium state between addition and removal of the gases might be reached some time in the future.

Many possible consequences have been assigned to this global warming, but there is no full agreement as to the possible magnitudes of these concomitant changes. Despite statements made by environmentalists and by 'bio-geo-politicians', attempts made during the preparation of this book to find any global connection between climatic change and the marine fish ecosystem have failed. However, regional climatic fluctuations have been common in the past (see Section 7.2).

Much attention has recently been devoted to the El Niño and La Niña phenomena in the Pacific, which are local in nature but are thought to have quasi-global effects. The real cause of these phenomena is still uncertain, as are their possible but often spurious connections with other parts of the world. It is, however, certain that a shift of the wind systems in the Pacific Ocean is involved, which affects mostly the upwelling along the Peruvian coast. A brief summary of the effects of El Niño is given in Chapter 3 section 2 and Chapter 7 section 2.

Another global phenomenon, the sunspot cycle (mean 11 years, variation 9 to 13 years) has been linked to various other quasi-cyclic climatic phenomena. Most of the links have failed, as no direct cause and effect relationship has been established (see further in this section and Fig. 7.8).

In studying the environmental changes in the oceans and their causes in the surface meteorological driving forces, the author is inclined to agree with Mörner (1984a, b) that although some climatic changes might have a global appearance, most changes are regional and/or hemispheric, and the changes are counter-balanced by compensatory changes with the opposite sign in other regions. Mörner maintained that climatic changes cannot originate from global rises and falls of temperature, but must represent redistribution of heat over the globe, locally, regionally or hemispherically. This distribution of heat also involves ocean circulation changes and possibly some changes of ocean heat content in intermediate depths, and so its time frame must be in the order of 50 to 150 years.

Robinson (1960) investigated possible long-term climatic change in the deep waters of the North and South Pacific Oceans. She found large, random individual differences down to 1000 m, but significant anomalies in several instances between 1000 and 5000 m. When the data were grouped by years or by water masses, scattered significant differences were observed at all levels and in all groups. However, she concluded that much larger samples are needed to separate statistically the short-period variability and the instrumental, processing and interpolation errors from long-period changes in deep waters.

Truly global climatic changes are insignificant, but considerable regional fluctuations of climate in historical times have occurred in medium and high latitudes, the largest well-documented fluctuations having occurred in the North Pacific and North Atlantic Oceans. If the minor quasi-global climatic fluctuations have had any effect on fish ecosystems, we need to seek these effects first in pelagic species which occupy the near-surface layers which are subjected to climatic changes.

The fluctuations of sardine and anchovy stocks show some global characteristics which have been assumed to be caused by large-scale climatic changes (Lluch-Belda *et al.* 1989). Stocks of both species in a number of regions show expansion of their range during high abundance and contraction during low abundance (Figs 7.1 to 7.4) and also some shifts of distribution along shore. These fluctuations are assumed to be associated with environmental changes of an oceanic scale. However, environmental changes which could explain these stock fluctuations, and would fit the space and time scales, have not been documented. Although regional environmental changes might have influenced the changes in some regions (see Chapter 5 Section 2), heavy fishing might have been the main cause of the collapse of several pelagic stocks. The timing of the peak catches and stock collapses is also considerably different in different regions so that the possible effect of ocean-wide climatic change might be ruled out.

Another characteristic of these neritic (pelagic) fish stock fluctuations is that when one species declines (e.g. sardine), it is replaced with another species (e.g. anchovy). Daan (1980) studied the possible replacement of depleted stocks by other species and concluded that very few reported profound fish ecosystem changes can be strictly classified as replacement phenomena. The California and South Africa sardine–anchovy species pairs came closest to replacement, as relatively simple direct food competition is

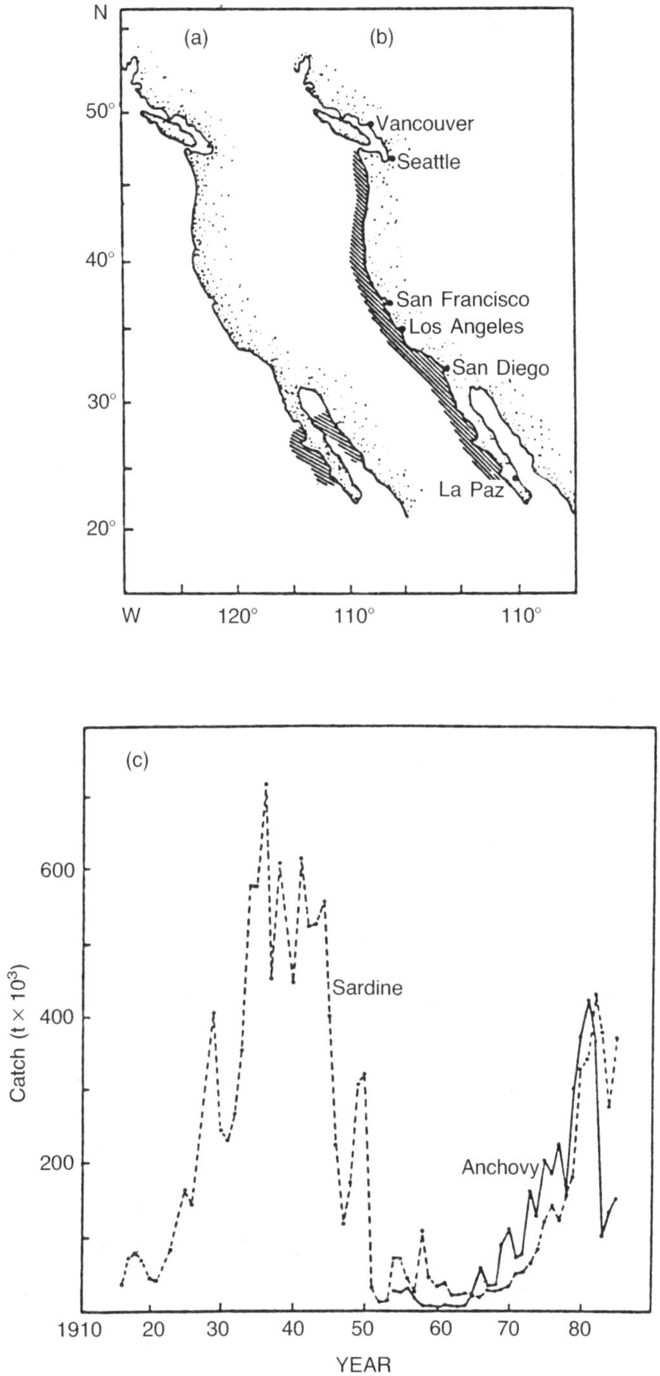

Fig. 7.1 Information pertaining to the sardine fishery of the California Current; distribution during periods of (a) low and (b) high abundance, and (c) trends in annual catches of sardine (dashed line) and anchovy (solid line) (Lluch-Belda *et al.* 1989).

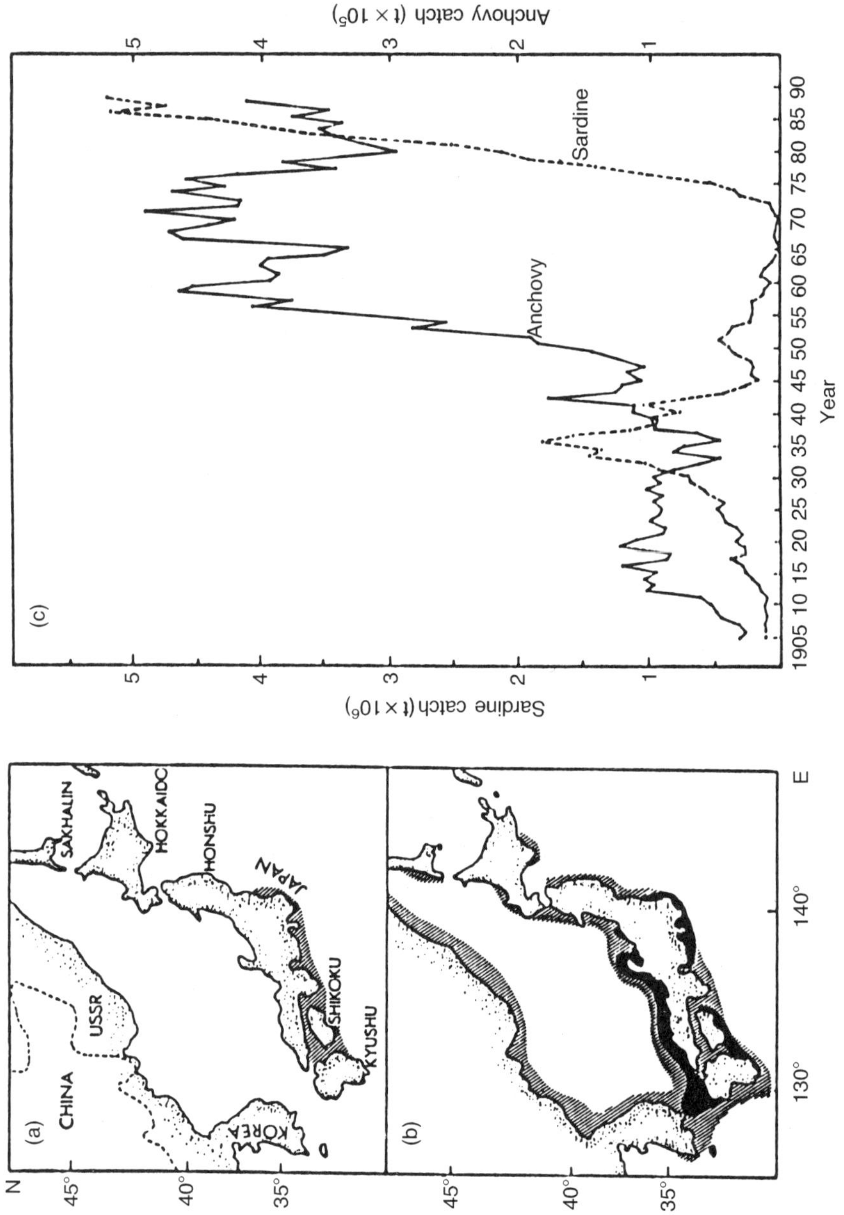

Fig. 7.2 Information pertaining to the Japanese sardine fishery; fishing (hatched) and spawning (black) grounds during periods of (a) low and (b) high abundance, and (c) trends in the annual catches of sardine (dashed line) and anchovy (solid line) (Lluch-Belda *et al.* 1989).

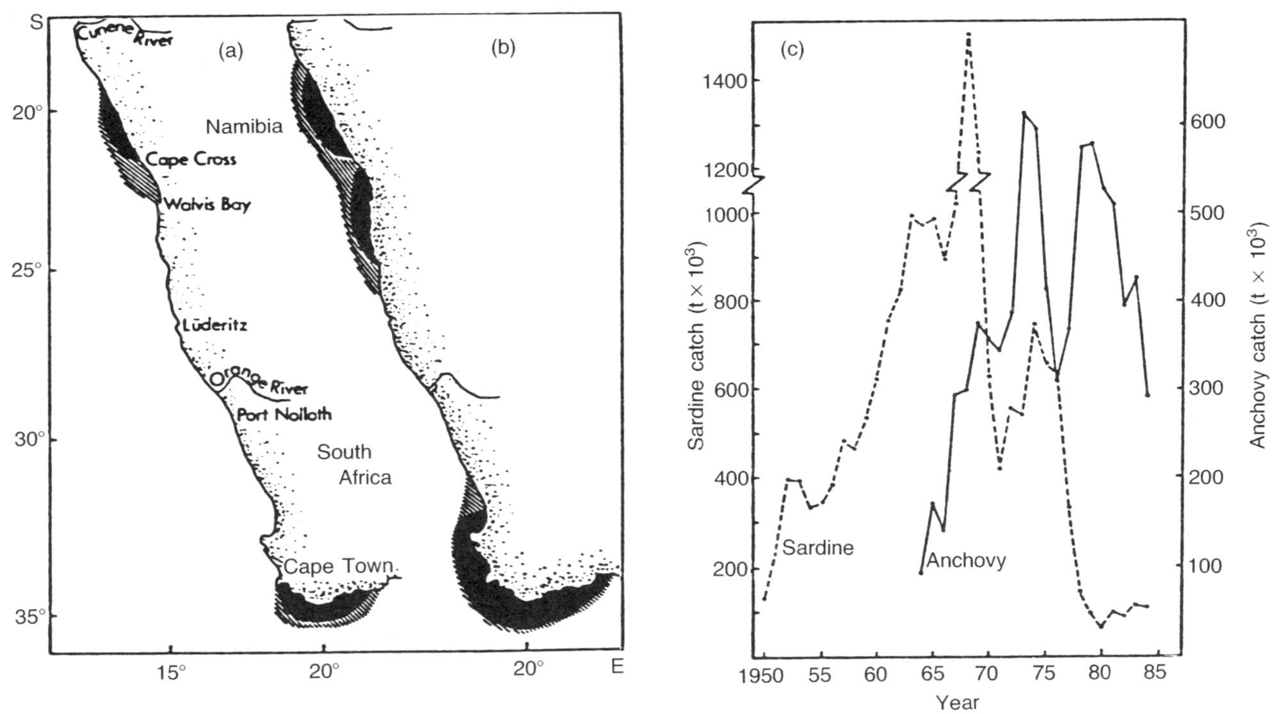

Fig. 7.3 Information pertaining to the sardine fishery of the Benguela system (based on information in Crawford *et al*. 1987); major fishing (hatched) and spawning (black) grounds during periods of (a) low and (b) high abundance, and (c) trends in annual catches of sardine (dashed line) and anchovy (solid line) (Lluch-Belda *et al*. 1989).

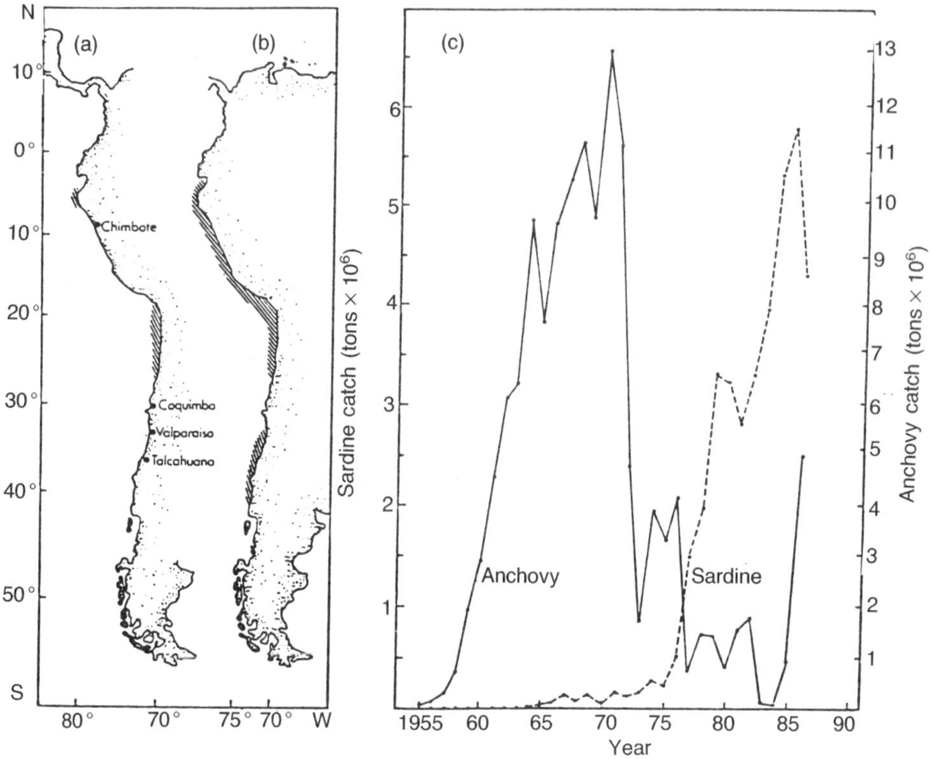

Fig. 7.4 Information pertaining to the sardine fishery of the Humboldt Current; spawning grounds during periods of (a) low and (b) high abundance and (c) trends in annual catches of sardine (dashed line) and anchovy (solid line) (Lluch-Belda *et al.* 1989).

involved and the trigger for the reported changes might have been partly of oceanographic origin with over-exploitation playing a secondary role. However, MacCall (1983) maintained that California sardines were depleted by extensive fishing pressure and were slow to recover. With some delay the anchovies increased and the stock is now fully exploited.

Thus no global or ocean-wide fluctuations of pelagic stocks can be found which might indicate the existence and effects of large-scale global climatic anomalies. Regional changes are considered in Chapter 7 section 2.

The large scale hydroclime fluctuations in the North Atlantic have been intensively monitored during this century (Dickson & Lamb 1972, Smed *et al.* 1983). Therefore, if large-scale, ocean-wide effects of climatic changes influence the fluctuations of fish stocks, the North Atlantic Ocean might be the best area to look. Of the large-scale fluctuations of pelagic stocks those of the Atlanto-Scandian herring seem to be most influenced by environmental anomalies (see Section 5.2). The drastic change of feeding migrations of the Atlanto-Scandian herring (Fig. 7.5) was mainly caused by environmental

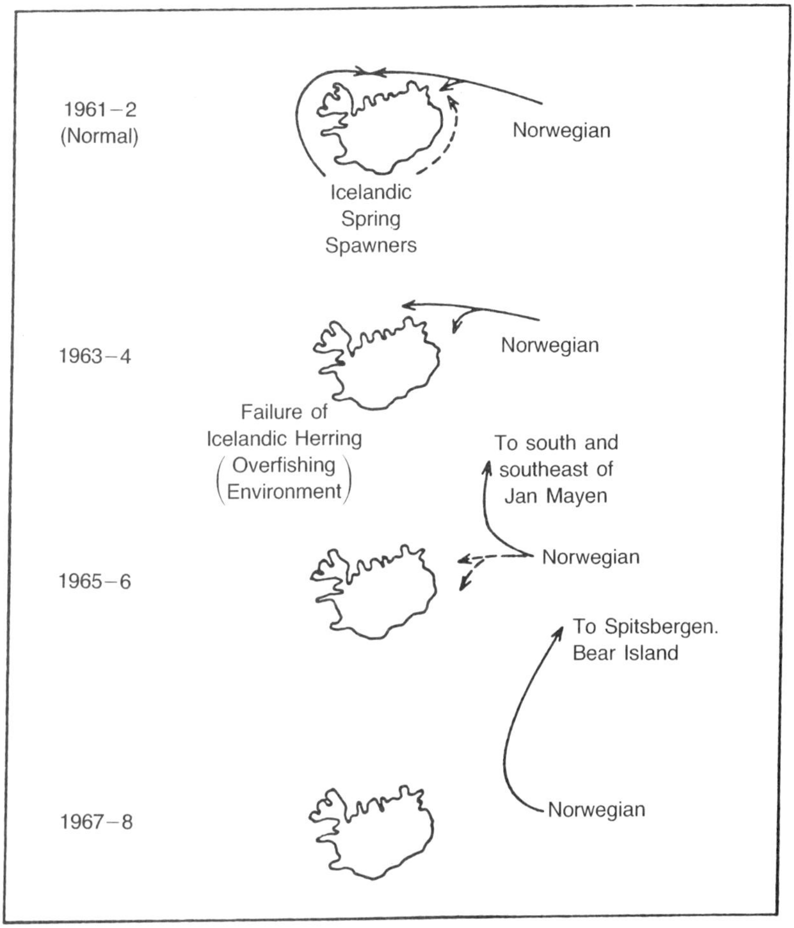

Fig. 7.5 Summary of recent changes in the feeding migration of herring at Iceland (Dickson & Lamb 1972).

changes north of Iceland (Jakobsson 1979). The herring stocks in the North Sea and the French sardine and herring stock fluctuations may also have been affected, but not caused, by local environmental conditions rather than by ocean-wide hydroclime changes.

Some semi-pelagic stocks may also be affected by global climatic fluctuations if and when they occur. The cod stocks in the northwest Atlantic have shown small variations until recently (Cushing 1986). From 1580 to 1750 annual catches of up to 300 000 t were probably taken and from 1750 to 1935 annual catches may well have reached a peak of 500 000 t. On the other hand, the fluctuations of the West Greenland cod stock may have been considerably affected by local (regional) climatic fluctuations (Cushing 1982b), and this is one of the few fish stocks where the effect of regional climatic fluctuation can be demonstrated (see Figs 5.17 and 5.18). The Arcto-Norwegian cod stock, although influenced in various ways by climatic fluctuations (Sundby & Sunnanå 1990), has shown a relatively steady decrease since the second World War under heavy fishing pressure (Fig. 7.6). Thus no common trend in cod stock fluctuations in the west and east North Atlantic stocks can be traced, and any effect of climatic change on these stocks is regional rather than global or ocean-wide.

The climatic fluctuations in the North Pacific, especially the movement of the mean position of the Aleutian Low, have been investigated by Favorite & Ingraham (1978). Mean pressure data from the winter half-years (October–March) of 3-year periods centred around the sunspot maxima, and corresponding periods centred around the minima, indicated a pronounced westward shift in the mean position of the Aleutian Low pressure system from the Gulf of Alaska to the Western Aleutian Island during years of sunspot maxima (Fig. 7.7). Wind-stress transport calculations indicate a

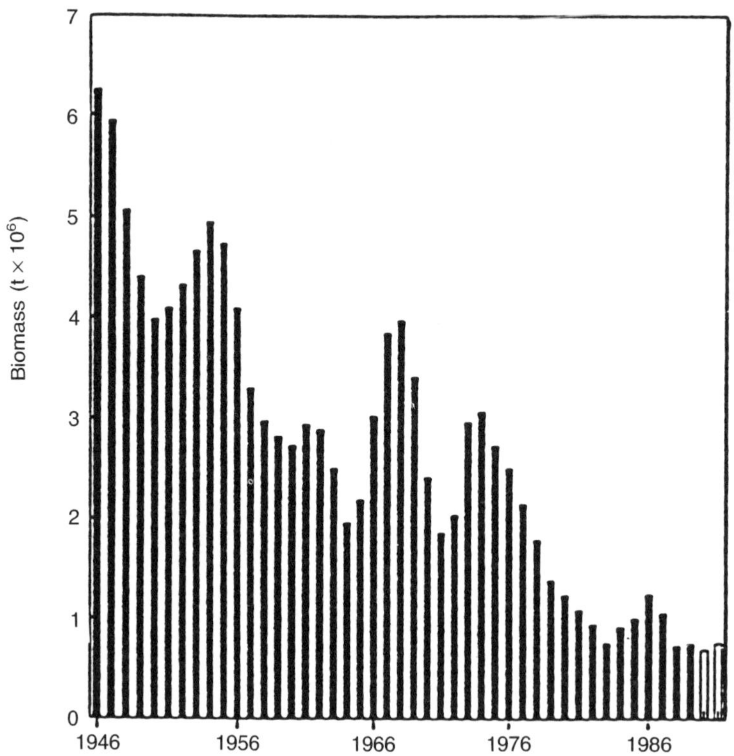

Fig. 7.6 Total biomass (millions of tonnes) of the Arcto-Norwegian cod stock since 1946 (Fisken og Havet 1990, Saernummer 1) (Sundby & Sunnanå 1990).

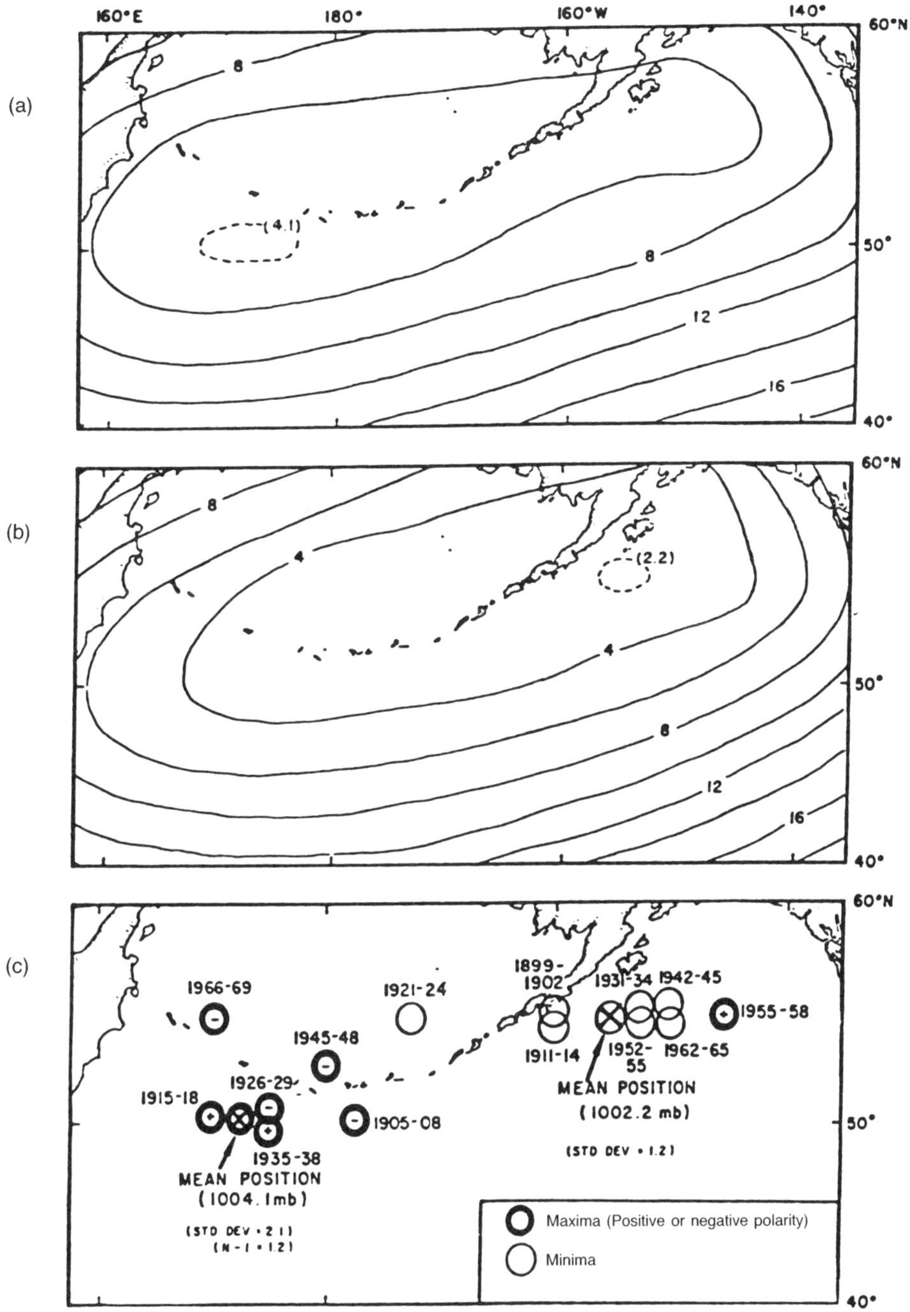

Fig. 7.7 Mean sea level pressure distributions (mb − 1000) for winter half-years (October–March) of 3-yr periods centred around sunspot maxima (a) and sunspot minima (b), and locations of centres of the Aleutian Low for individual periods (c), 1899–1974 (Favorite & Ingraham 1978).

20% reduction in northward transport into the Gulf during periods of sunspot maxima compared with that during sunspot minima. Unfortunately, any teleconnections or servomechanisms between sunspot activity and physical or biological phenomena on earth are not clear at present.

In attempts to determine the possible relationships between long-term environmental changes (which are usually semi-cyclic with average periods of 6 to 23 years), and the changes in the fish ecosystem in the same time interval, it becomes clear that the fishery data from the northeast Pacific are poor in time and space coverage as well as in quality. In contrast, fishery data from the North Atlantic, where there is a long history of intensive fishery research, are relatively good. Moderately reliable age composition data, which would allow the determination of good or bad recruitment, can be found in the northeast Pacific for only a few species and for a relatively short recent period. The reported determination of stock sizes and their fluctuations is entirely unreliable and mixed with personal opinions and various distortions of data.

The only species from the northeast Pacific on which longer records are available is the Pacific salmon, which roams around in the North Pacific. The catch returns vary in space and time (see Fig. 7.8), and are affected by numerous local factors, such as conditions of smolt production in fresh water, enhancements, fishery regulations, variable oceanic catches, etc., so that any effect of climatic change on stock fluctuations is impossible to detect.

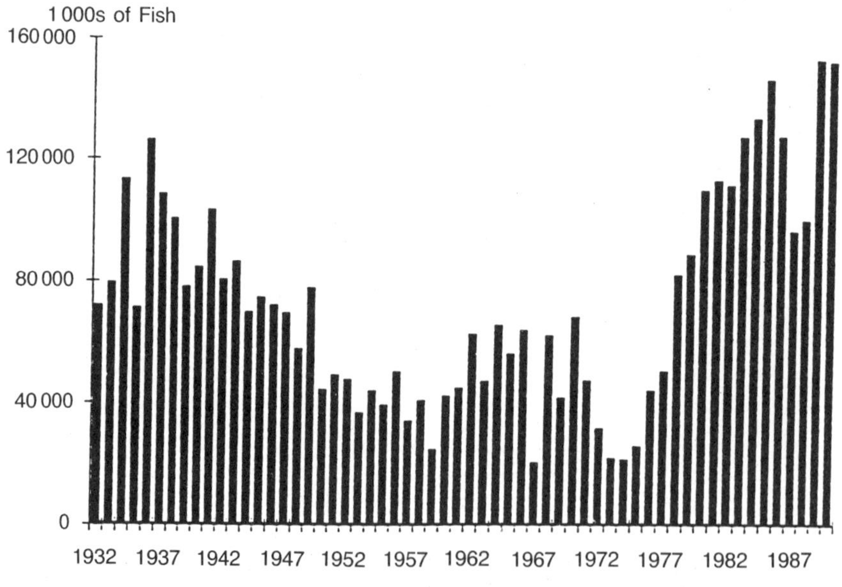

Fig. 7.8 Alaska salmon catch, 1932–1986, in thousands of fish (Alverson 1991).

Thus attempts to find global and/or ocean-wide fluctuations of fish stocks which might have been at least partly either initiated and/or caused by these large-scale climatic fluctuations proved unsuccessful. However, several regional profound fluctuations were shown, especially in pelagic species, which might have been affected by local climate and/or hydroclime fluctuations. These are described in Chapter 7 section 2.

7.2 Regional and coastal changes in fisheries

In Chapter 6 a number of examples of local fish stock fluctuations have been given, together with brief notes on the possible effect of climatic changes. Cushing (1982a) described additional historic sporadic appearances and disappearances of fishes in some locations. Whether these local appearances and disappearances were caused by changes in stock size or in migration routes, or were affected by environmental factors or other causes, cannot always be determined with certainty. The various attempts to correlate historic periods of high catches of different herring stocks in the North Atlantic with cold and/or warm climatic periods and with the extent of ice cover in the Baltic and north of Iceland, have been described by Cushing (1982a). Simple correlations are not convincing, especially if the data quality is questionable and cause–effect relationships cannot be explained. In most cases climatic and fish catch data are linked in statistical fashion without any attempt to clarify functional relationships (cause and effect). Simple correlations tend to break down when more information becomes available.

Some regional hydroclime fluctuations and catches (or landings) of major commercial species are summarized in this section. Attempts are also made to find whether local climatic changes might have affected these fluctuations.

North Atlantic

Surface climatological and hydroclime fluctuations in the North Atlantic have been analysed and summarized by many scientists (e.g. Tait (1953), Dickson (1968), Martin (1969), Malmberg & Svansson (1982), Ellett & Blindheim (1991), Malmberg & Kristmannsson (1991) and others). Analyses have been made of various weather and oceanographic elements on many space and time scales. Some selected results of these fluctuation studies are given below, emphasizing the particular hypotheses of the authors.

Surface temperature is not a conservative property. It changes seasonally and is a function of surface winds (advection and mixing) and local heat exchange. Salinity is, however, more conservative. Already before the discovery of the advection of the Great Salinity Anomaly of 1970s Martin (1969) had found that salinity is associated with decrease in the Gulf Stream flow (Fig. 7.9) to a stronger degree than temperature. His findings gave

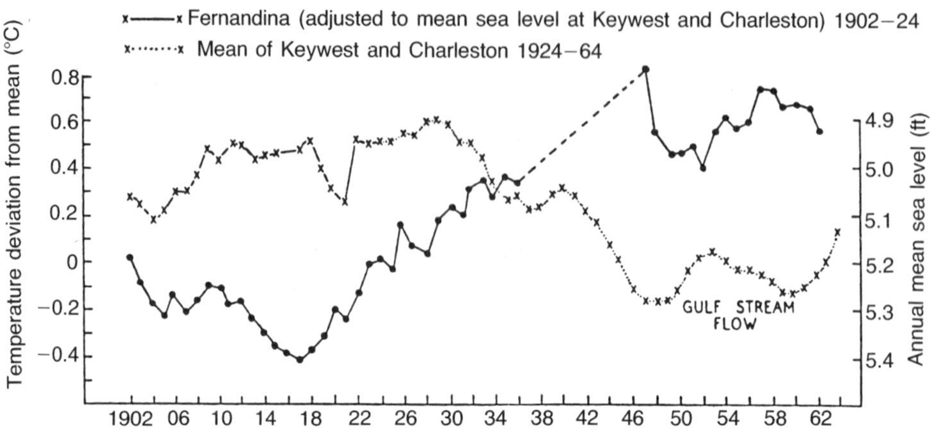

Fig. 7.9 Five-year running means of surface winter sea temperature in sea area (J + K + L) and of the Gulf Stream Flow (Florida Current) (Martin 1969).

additional support to Iselin's hypothesis that thc major flow of warm saline water out of the Gulf Stream towards the Norwegian Sea would occur when the strength of the Gulf Stream system was declining.

Dickson (1968) in analysing 60-year salinity records (1905 to 1965) for the shelf seas of the North Atlantic, found that fairly regular (but not cyclic) variations occur between periods of low and high salinity which are only roughly contemporaneous events. He found that a plausible cause is advection, caused by large-scale surface wind shifts (circulation anomalies) (Fig. 1.4).

The selection of environmental data for the study of possible effects of environment on fish stocks is best done with a defensible hypothesis at hand so that the selected data are pertinent to their possible effect on the species and/or its life process. For example, the selection of the environmental variable in respect to depth and its treatment, e.g. averaging in area and time, can show considerable differences. This is illustrated in Fig. 7.10, which shows that the variation of surface and 30 m temperature and salinity, as well as the average, all present a different picture, and can be meaningless in the sense that fish can scarcely integrate the variation of temperature or salinity over 5 or 17 years!

The studies of climate and hydroclime fluctuations in the decade 1980 to 1990 in the North Atlantic and in the North Sea (Ellett & Blindheim (1991), Malmberg & Kristmannsson (1991)) indicate that the decade was normal, the surface salinity recovering from the Great Salinity Anomaly. Although this decade was normal in regard to the environment, large changes occurred in fish stocks which were summarized by Jakobsson (1991) (see later this section).

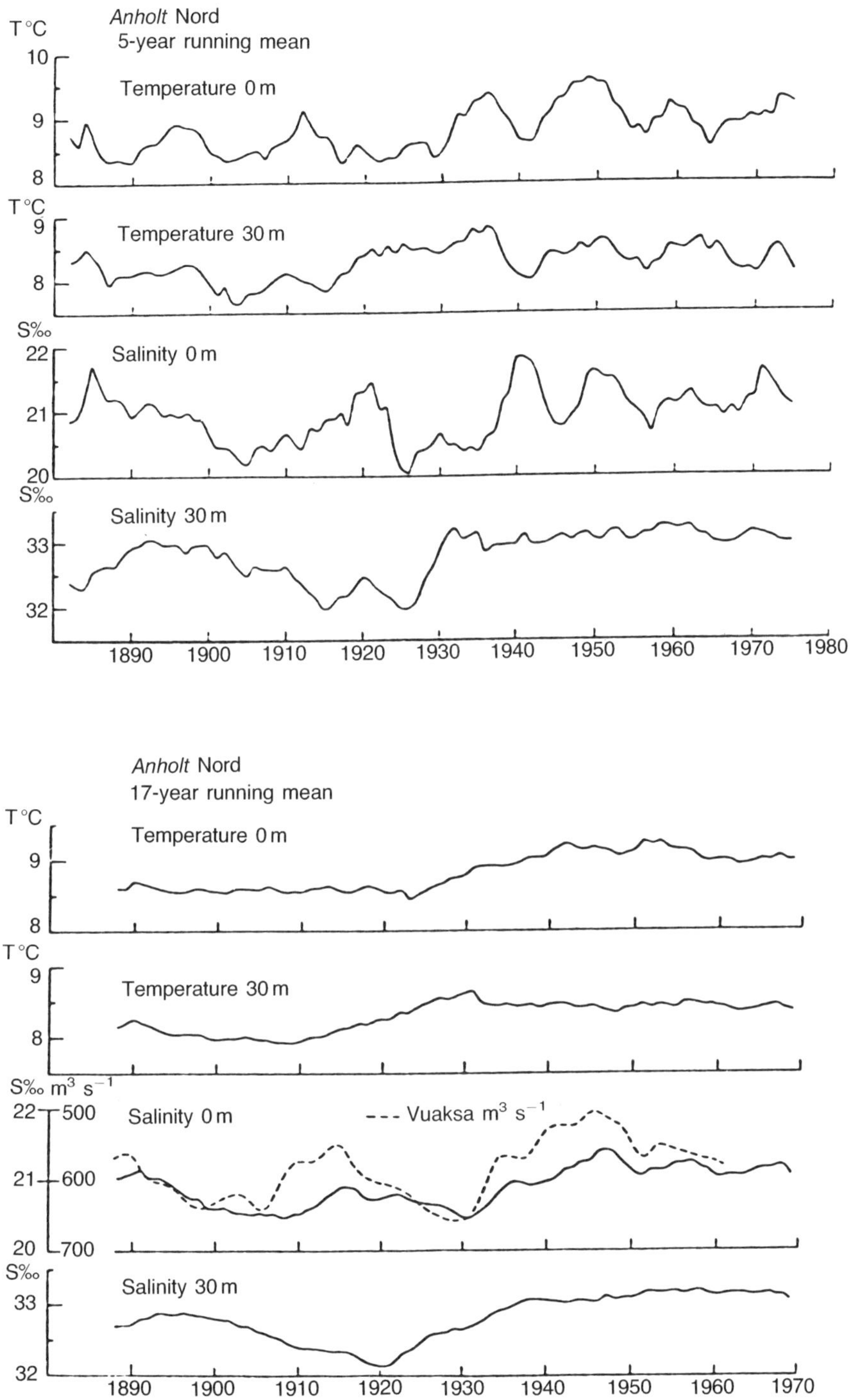

Fig. 7.10 Five- and seventeen-year running means of temperature and salinity during the period 1880 to 1977 at the Danish light vessel *Anholt* in the Kattegat (Malmberg & Svansson 1982).

The greatest changes in fish stocks in the North Atlantic in last 40 years have occurred in the herring stocks (Figs 7.11 to 7.13) (Bakken 1983). There seem to have been several causes for these changes involving both fishery and environment. The heavy fishing on the adult stock as well as on the juveniles is most often pointed out as a plausible cause of herring stock collapses. However, the migrations of Norwegian spring-spawning herring also changed. Jakobsson (1980) summarized some of these changes and described their complexity:

> 'The geographical distribution of the catches of herring in the 1960 decade in Icelandic waters changed drastically. In 1960–1962 the catches were taken at the north and east coast of Iceland, in 1963 the catches off the north coast were negligible while some catches were taken at Jan Mayen. In 1964 practically all the catches were taken at the east coast of Iceland. In 1965 and 1966 the catches were distributed over wide areas stretching from Jan Mayen to the southeastern border of the cold Icelandic current. In 1967 and 1968 the summer fishery shifted once more and took place on fishing grounds at Bear Isle and Spitsbergen.
>
> The years 1960–1964 were considered to belong to the series of 'normal' years comparable to the period 1950–1960 although considerable variations were observed. In late summer 1964 indications of a different hydrographical regime were observed and in 1965, 1967 and 1968 the increased flow of Polar water in the area north and east of Iceland caused a sharp decrease in temperature in this area. Moreover, the traditional herring grounds north of Iceland were more or less covered with ice.
>
> This drastic change in the hydrography was associated with a sharp decline in primary production of phytoplankton during the ice years. At the same time a collapse of the often abundant stock of *C. finmarchicus* was observed. Its depletion was so severe that it did not show signs of recovery until in the mid seventies.
>
> Taking into account that the north and east Icelandic coast herring fishery was based on feeding migrations it is evident that the adverse environmental conditions which prevailed north and east of Iceland in the late 1960s played a major role in changing the migration pattern of the Atlanto-Scandian herring which in turn caused the termination of the herring fishery north and east of Iceland.'

Jakobsson (1980) also described the collapse of Icelandic spring and summer spawning herring as caused by heavy fishing. The summer spawners have showed a rapid recovery (Fig. 7.13) whereas spring spawners have not.

The collapse of the cod fishery in West Greenland has been described in Chapter 5 section 3. The West Greenland cod stock has not recovered, despite temperature having been above the average (Fig. 7.14). It could be

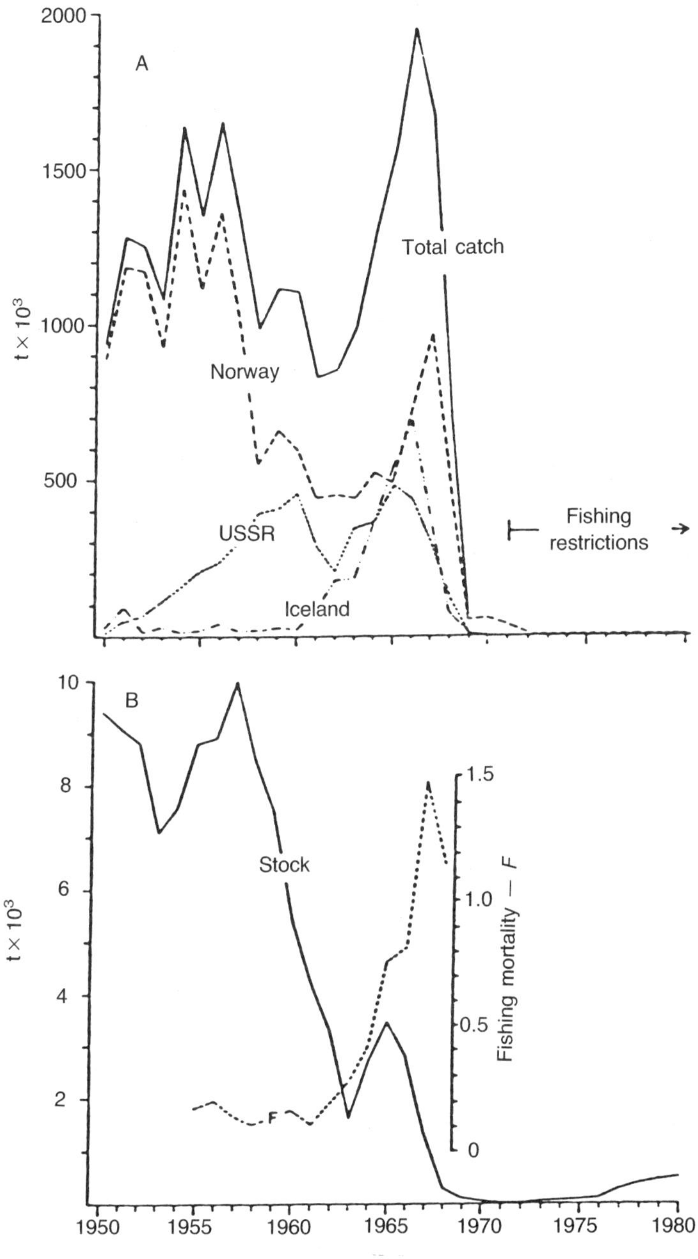

Fig. 7.11 Norwegian spring spawning herring. A. Annual catch, total and by major fishing nations, demonstrating the increased catch in the late 1960s and the collapse of the fishery by 1970. B. Biomass of spawning stock (age >5 years) and fishing mortality rate (age >7 years) 1955–68, showing the stock decline and increased exploitation during the 1960s. (Data and figure adapted from Dragesund (Bakken 1983).

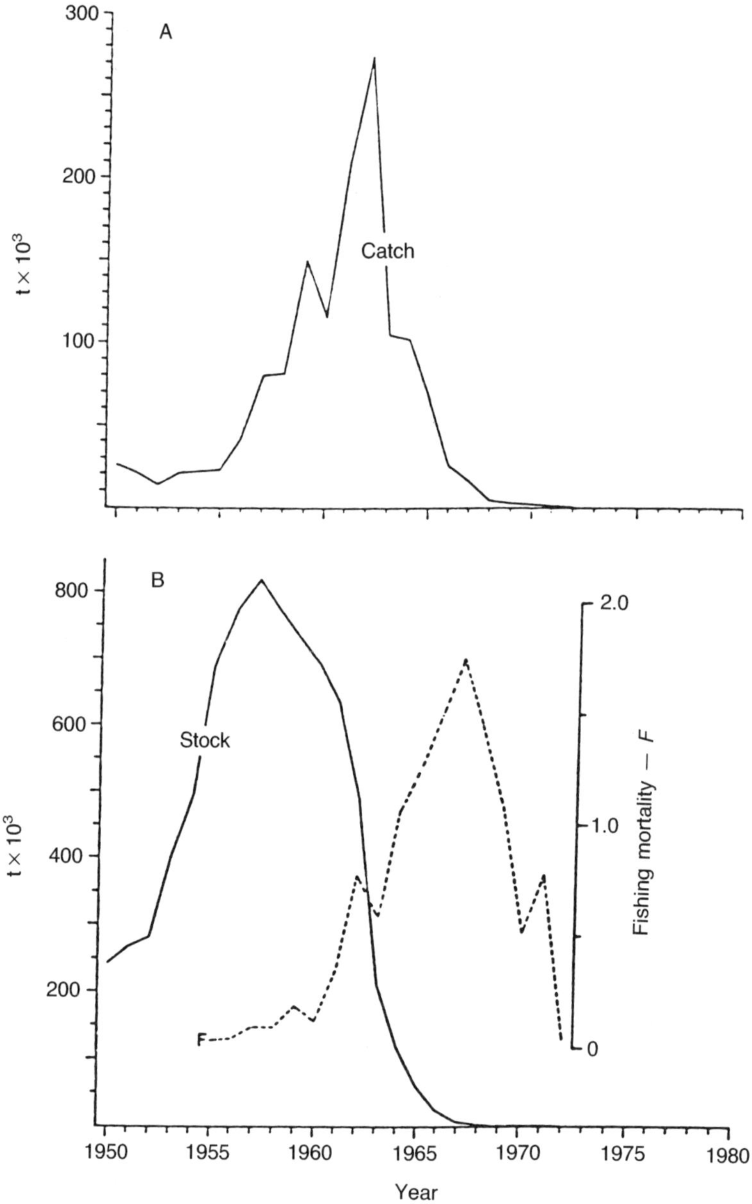

Fig. 7.12 Icelandic spring-spawning herring. A. Annual catch. B. Biomass of spawning stock (age >4 years) and fishing mortality (weighted mean, age 4–15 years) 1955–72 illustrating the relationship between stock decline and fishing mortality increase (Bakken 1983).

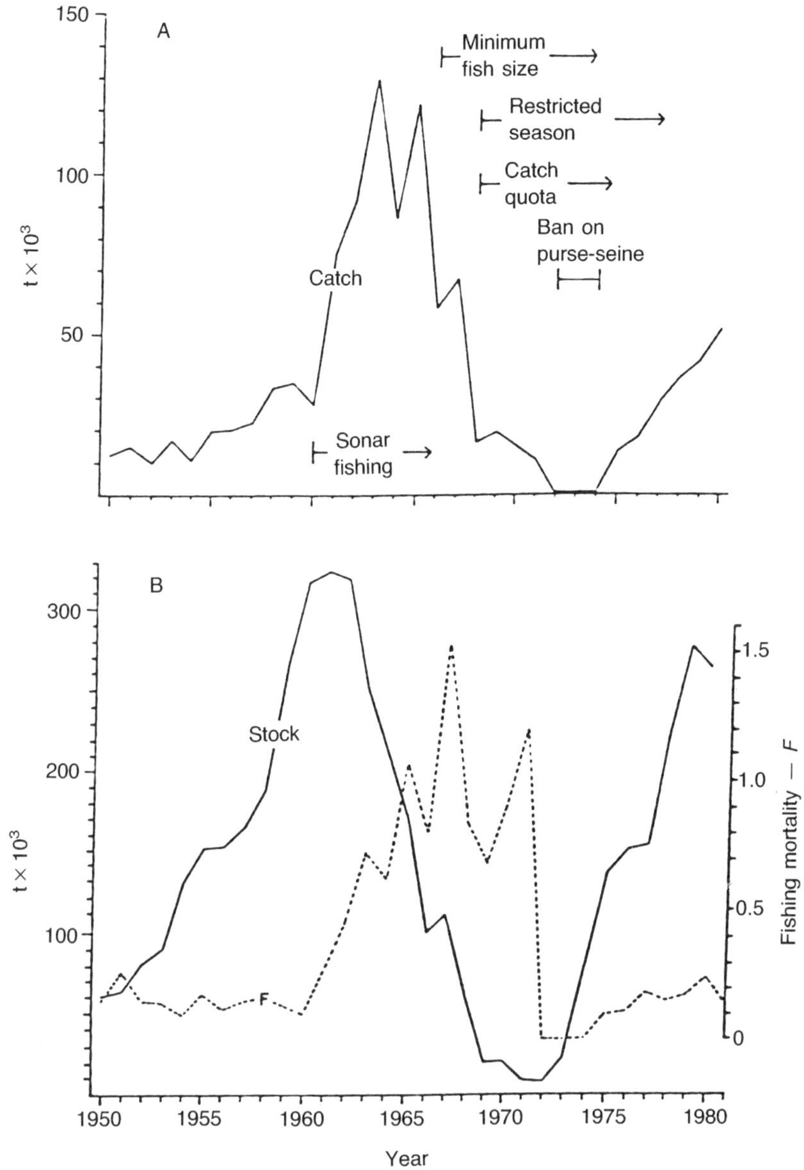

Fig. 7.13 Icelandic summer-spawning herring. A. Annual catch showing the fluctuating yield 1955–1980. B. Biomass of spawning stock (age >3 years at the time of spawning) and fishing mortalities (weighted mean, 1950–59 age 3–10, later age 3–13) reflecting the effects of the unrestricted fishery and the fisheries management of the 1970s (Bakken 1983).

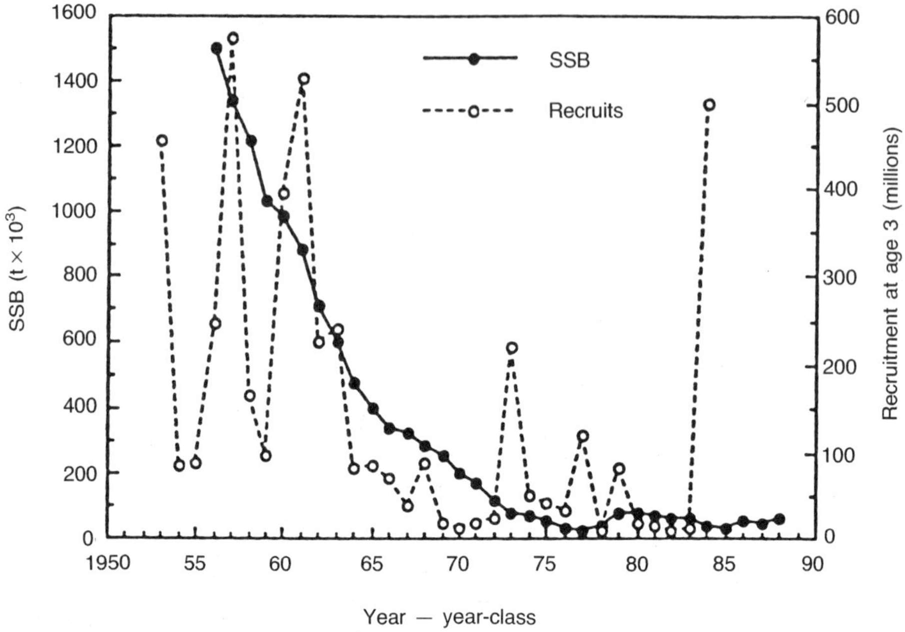

Fig. 7.14 Spawning stock biomass and recruitment in West Greenland cod, 1953–88 (Jakobsson 1991).

added to the earlier description in Chapter 5 section 3 that the West Greenland cod stock is dependent on the drift of Ø-group cod from Iceland to Greenland. These cod return to spawn in Icelandic waters, as happened with the 1973 year class which returned in 1980–1 (Jakobsson 1991). Good year classes of Icelandic cod are produced during the early warm period or just before its beginning — the same phenomenon as in Northeast Arctic cod in the Barents Sea, associated with the increased inflow of warm Atlantic water into both areas (Jakobsson 1991). Thus the West Greenland cod recruitment is apparently dependent on surface wind anomalies in Greenland-Iceland area.

The most dramatic event in the Icelandic fisheries in recent years was the collapse of the capelin fishery and its rapid recovery in the 1980s (Fig. 7.15). The collapse of capelin stock was associated with about 25% reduction of growth rate of 3- to 6-year-old cod (Jakobsson 1991). The reason for the collapse of capelin stock was poor recruitment during the period 1978 to 1981 which was associated with low zooplankton standing stock north of Iceland. The abundant 1983 year-class brought rapid recovery of the capelin stock.

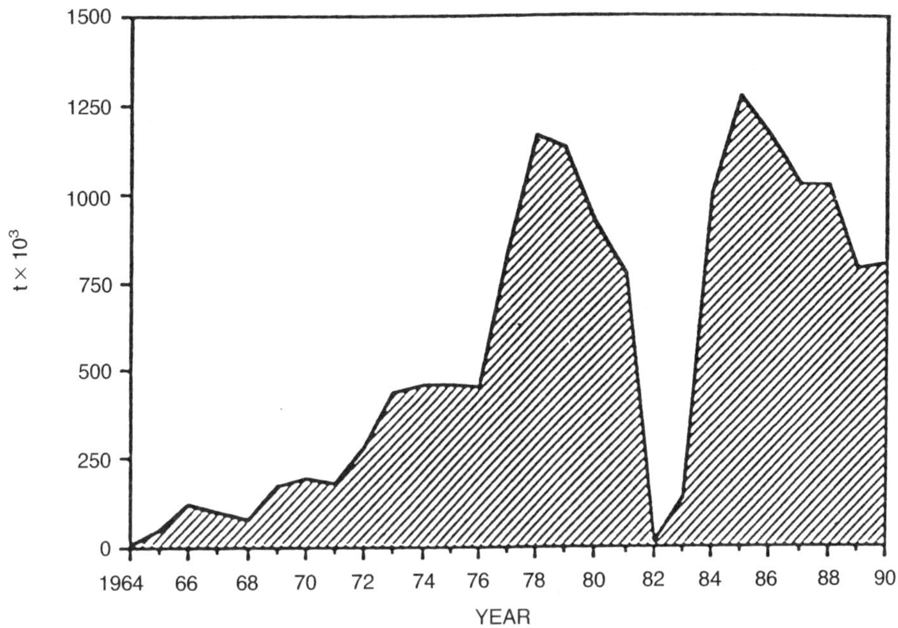

Fig. 7.15 Annual catches of Icelandic capelin, 1964–90 (Jakobsson 1991).

Barents Sea

Barents Sea is rich in fish, owing to high productivity, but poor in number of species. Most abundant fish species occurring there are close to their northern distribution boundary, and so they might be sensitive to climatic fluctuations. Furthermore, the area is heavily fished, and an interaction between fishing, climatic fluctuations and competition for basic food resources occurs there. In addition, some abundant fish stocks spawn outside the Barents Sea, along the Norwegian coast (cod, herring), and the larvae and juveniles must be carried into the Barents Sea with currents. The temperature fluctuations in the Barents Sea and along the Norwegian coast are also of an advective nature (Fig. 2.22) (Blindheim *et al.* 1981, Saetersdal & Loeng 1983).

The most important commercial species in the Barents Sea is the Northeast Arctic cod, followed by capelin and herring in earlier years. Capelin and herring are the main food items of adult cod. The fluctuations and collapse of the Atlanto-Scandian herring have been described earlier (see Fig. 7.11). The capelin stock collapsed in 1986 (Fig. 7.16), whereas herring showed some recovery at the beginning of the 1980s. The rapidly growing cod stock grazed down shrimp and the rest of the herring and capelin stocks in the

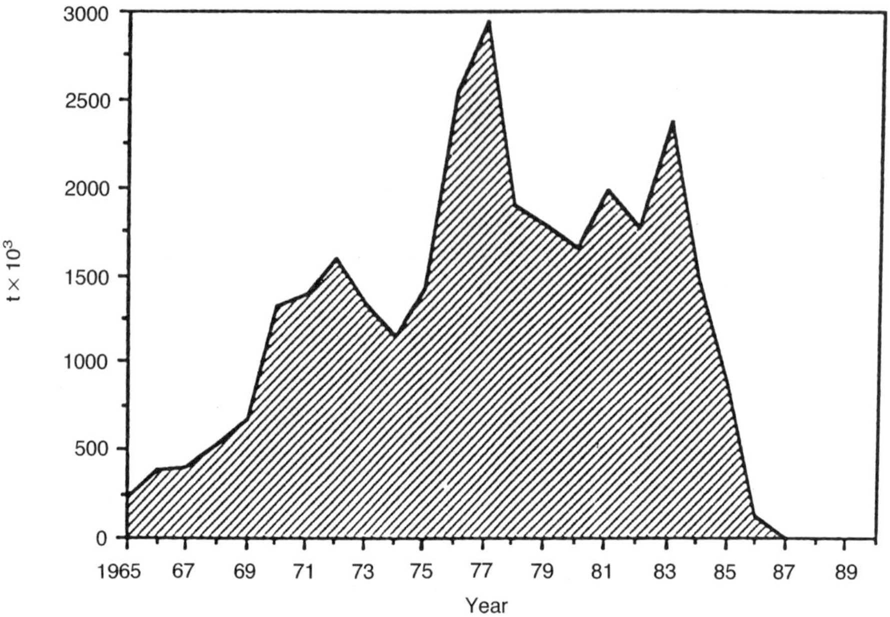

Fig. 7.16 Annual catches of Barents Sea capelin, 1965–90 (Jakobsson 1991).

Barents Sea. The collapse of the Norwegian herring stock in the 1960s coincided with a huge increase in the capelin population in the Barents Sea. Young herring are also potential predators on capelin larvae (Moksness & Øiestad 1987). This interaction is furthermore made more complex by the different spatial distribution of capelin in warm and cold years (Fig. 7.17) (Loeng 1989).

The Northeast Arctic cod has yielded about 780 000 t annually, with peak catches over 1 000 000 t (Fig. 7.18). In the 1980s the fishing mortality was high in relation to the stock size and the low catches then were due to a series of small year-classes produced during a cold period from 1976 to 1982 (Fig. 7.19). The collapse of the capelin stock had, however, a disastrous effect on the cod stock. It resulted in starvation of cod, the average weight being reduced by up to 50%, and even forced hungry Arctic seals to migrate down to the Norwegian coast to find food. The collapse of the Barents Sea ecosystem might well be related to, and/or initiated by, the collapse of the Atlanto-Scandian herring.

The production of good year-classes in the Barents Sea has been linked to the temperature regime (Saetersdal & Loeng 1983), and especially to proper feeding conditions (plankton production) for larvae and juveniles. Also the distribution of fish (cod, haddock and capelin) is more easterly and northerly in warm years than in cold years. Thus the regional climatic fluctuations affect

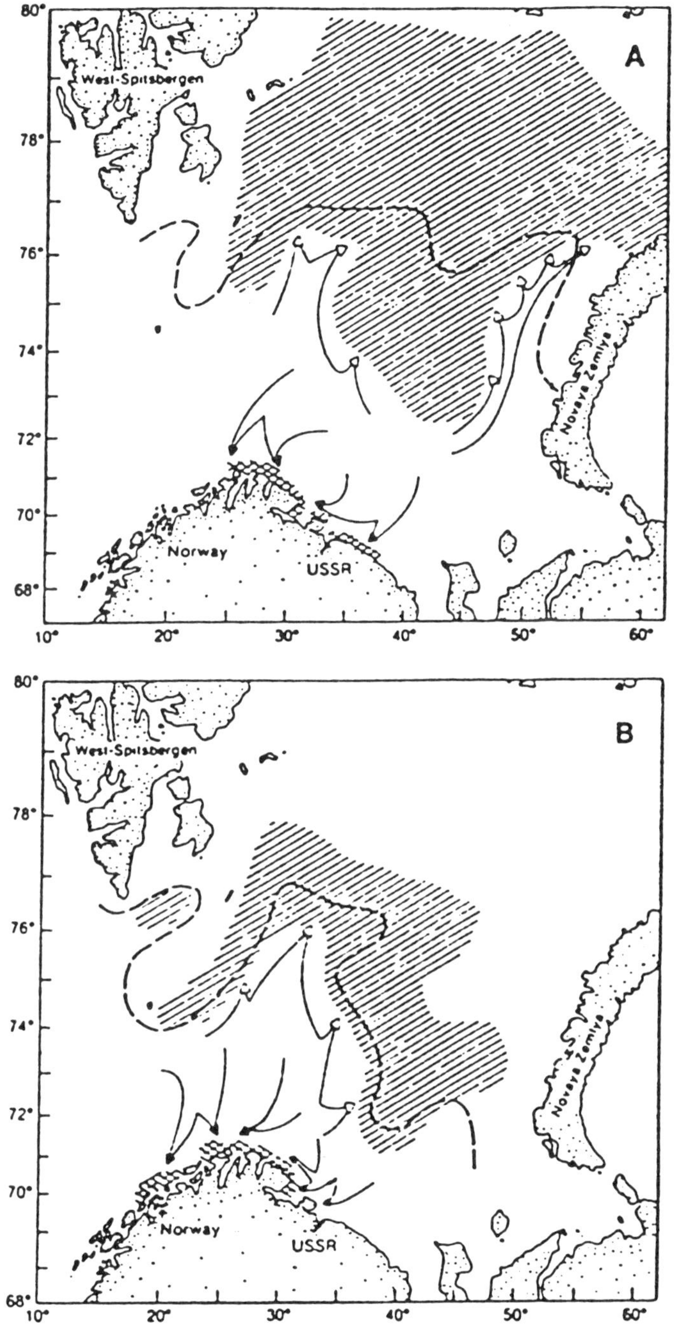

Fig. 7.17 Feeding distribution (hatched) and spawning ground (double hatched) of capelin in (A) warm years and (B) cold years. Open arrows indicate feeding migration; closed arrows indicate spawning migration routes. The dashed line indicates the approximate position of the 0 °C-isotherm at 100 m in warm and cold years (Loeng 1989).

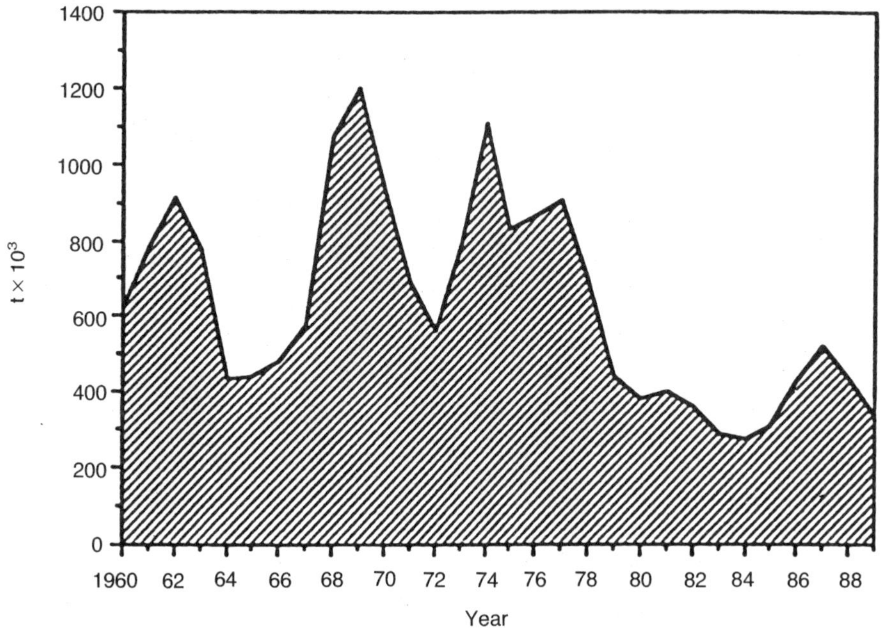

Fig. 7.18 Annual landings of the Northeast Arctic cod, 1960–89 (Jakobsson 1991).

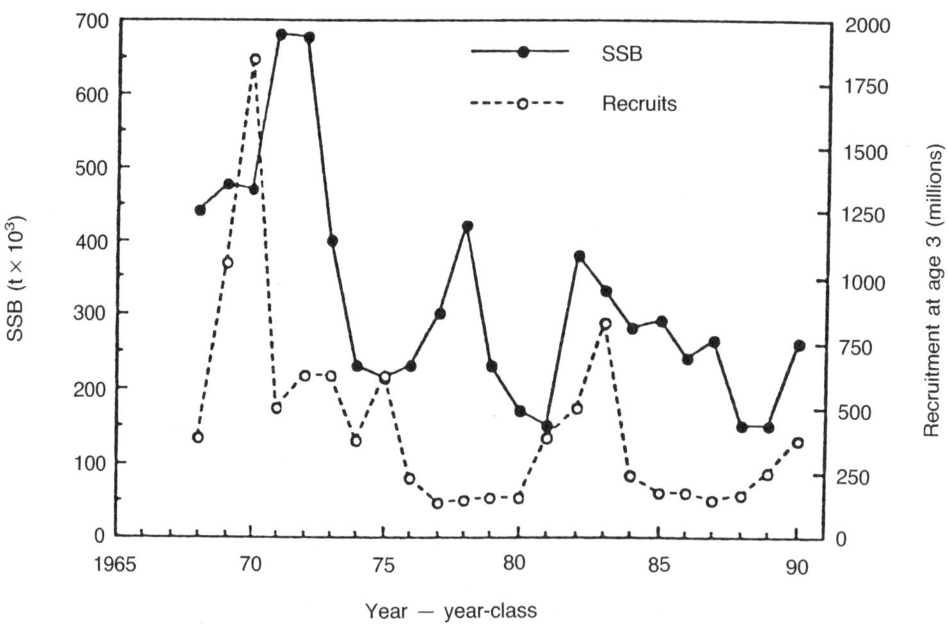

Fig. 7.19 Spawning stock biomass and recruitment in Northeast Arctic cod, 1968–90 (Jakobsson 1991).

marine resources in the Barents Sea to a large extent. However, the climatic effects are complex, affecting different species differently through effects on recruitment, larval survival, predation, and growth. The main influencing factor seems to be surface wind causing advectional changes in the surface layers.

Northwest Atlantic

In the Northwest Atlantic, including Georges Bank and the Grand Banks, there have been great fluctuations in catches, which are different in demersal and pelagic species (Fig. 7.20) (Brown *et al.* 1983). It is difficult to suggest any biological hypothesis on the demersal–pelagic interaction. Nor can the fluctuations be fully explained on the basis of fishing pressure, as fishing mortality rate for the principal groundfish species has been higher in the North Sea than in the Northwest Atlantic (Brown *et al.* 1983). Attempts to find long-term environmental correlations have not yet proved very successful.

Some profound fish ecosystem changes have occurred in the Northwest Atlantic which are not fully reflected in catches. In the late 1970s sand lance stock increased 50-fold after the decline of the herring and mackerel stocks, and in the mid-1980s the mackerel stock increased rapidly (Fig. 7.21).

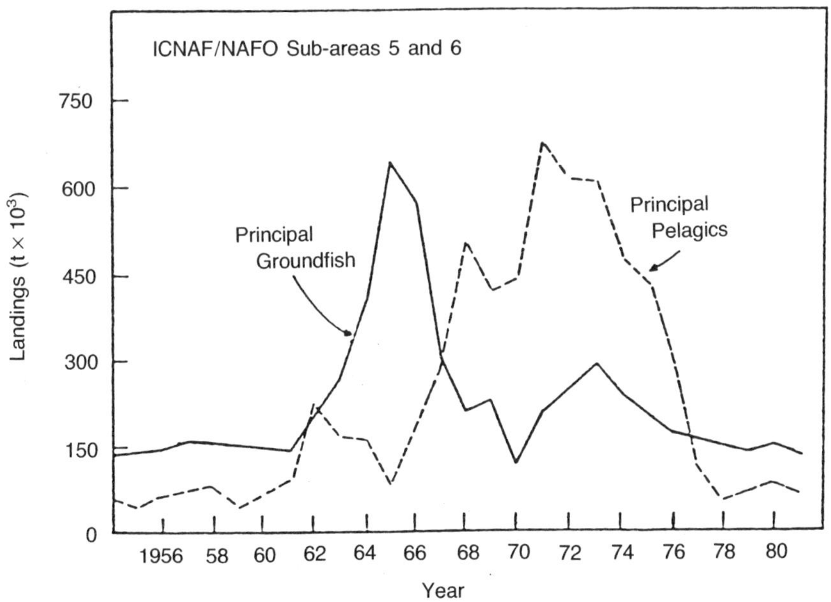

Fig. 7.20 Nominal catch of principal groundfish and principal pelagics in ICNAF/NAFO Sub-areas 5 and 6 (Brown *et al.* 1983).

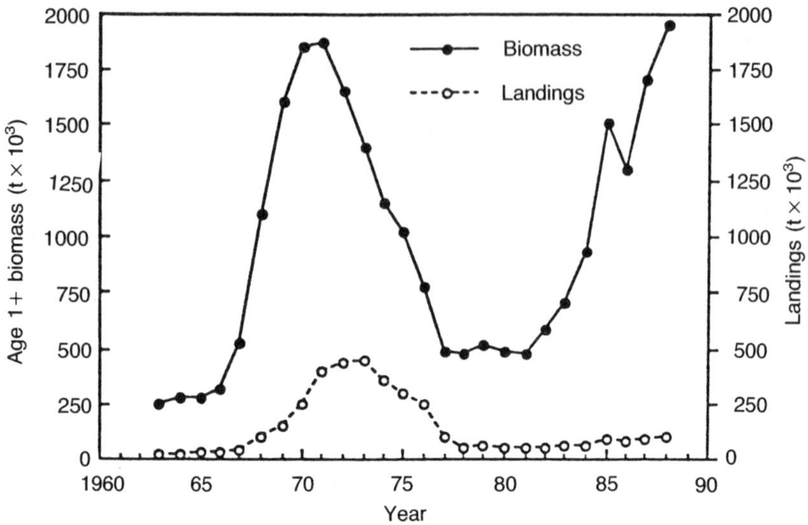

Fig. 7.21 Stock abundance and landings of Atlantic mackerel in the area of Labrador to North Carolina, 1963–88 (Jakobsson 1991).

The landings of demersal species from the Northeast coast of the USA have fluctuated considerably (Fig. 7.22). Two species contribute most to these changes, haddock, which has declined, and cod, which has increased. Although the spawning stock biomass of Georges Bank cod is low, the recruitment has not suffered. On the other hand, there have been failures in haddock recruitment (Fig. 7.23). Owing to heavy fishing pressure any possible effect of environmental or climatic variations on the stocks of haddock and cod are obscured.

In Canadian waters there are several cod stocks, the largest being the northern one (Fig. 7.24). The increased fishing pressure in late 1960 decreased the spawning biomass of northern cod and the recruitment remained at a low level. This cod stock is assumed to be the first stock of cod in the world which has been recruit-overfished (Jakobsson 1991). There have also been indications that large-scale meteorological patterns affected the recruitment of this cod stock (Koslow *et al.* 1987), but the mechanism of this interaction is poorly understood. On the other hand, it has been suggested that high recruitment of northern cod in the 1960s might be interpreted as a special short-term outburst of this stock due to some favourable environmental conditions (Jakobsson 1991).

The most noticeable happenings in the northwest Atlantic in the 1980s have been the rapid increase of the mackerel stock and the decrease of northern cod. Neither can be described as being directly determined or materially affected by climatic changes.

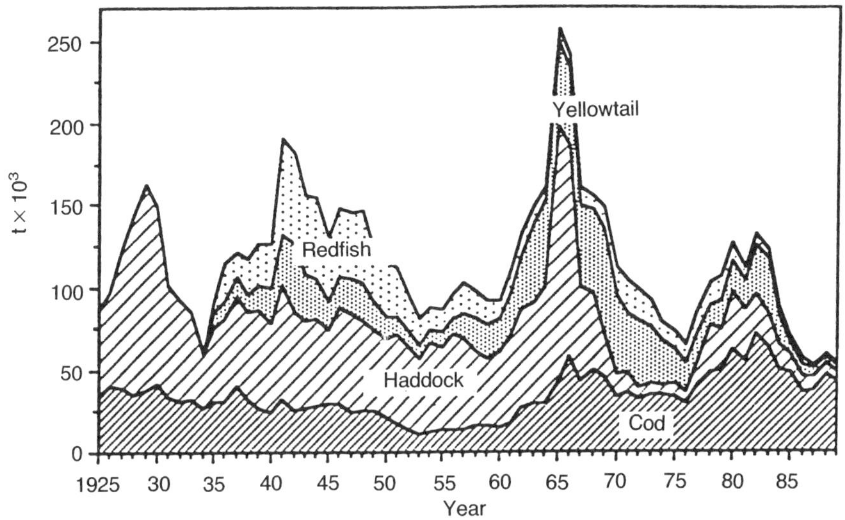

Fig. 7.22 Annual total landings of cod, haddock, yellowtail flounder and redfish from off the northeast corner of the USA, 1925–89 (Jakobsson 1991).

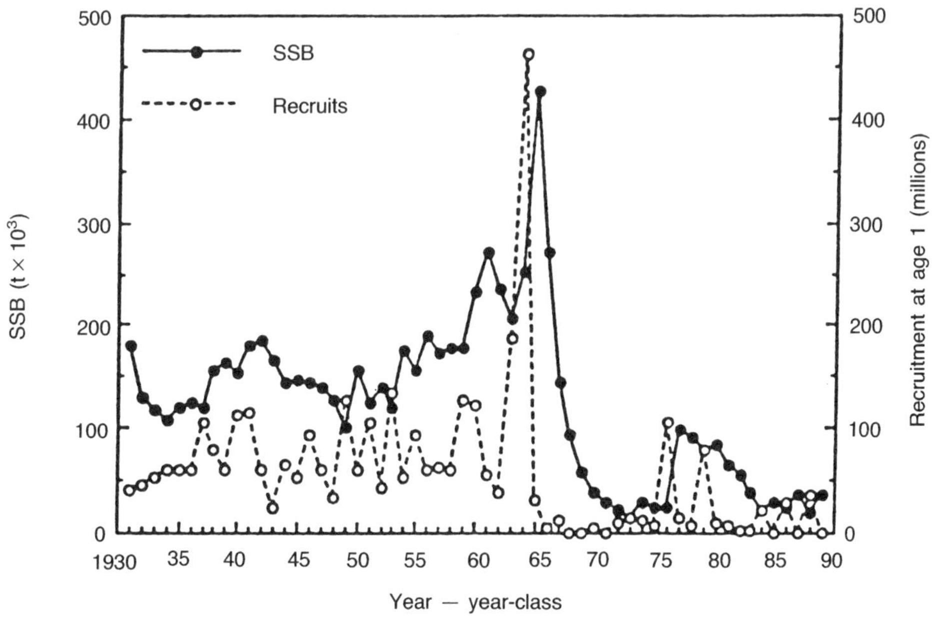

Fig. 7.23 Spawning stock biomass and recruitment in Georges Bank haddock, 1931–89 (Jakobsson 1991).

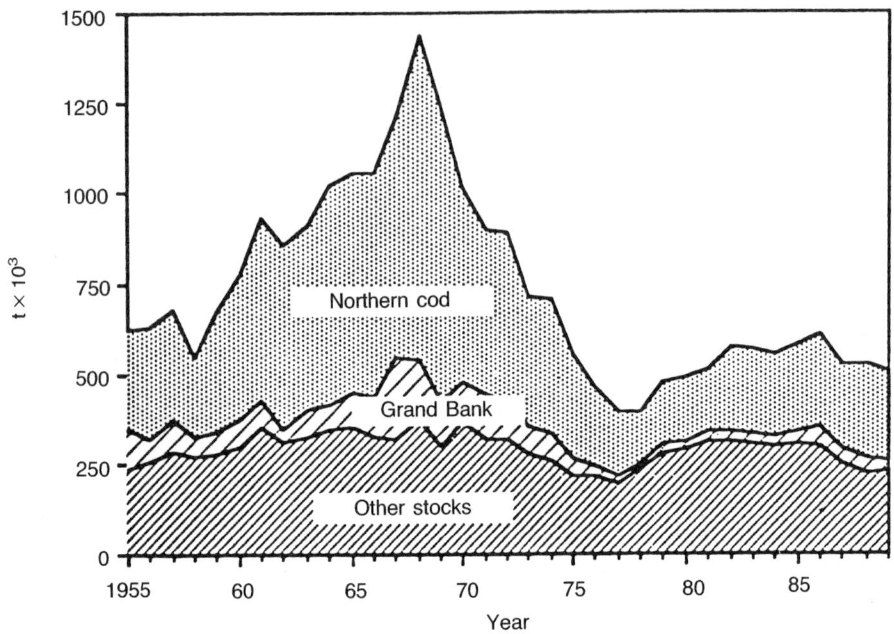

Fig. 7.24 Annual landings from Canadian cod stocks, 1955–89 (Jakobsson 1991).

North Sea

The changes in fish stocks in the North Sea have been summarized by Hempel (1978b) (see Chapter 5 sections 2 and 3). The most significant event of the decline of herring stocks and the gadoid outburst (Cushing 1980) is summarized in Chapter 5 Section 3 and the recent spatial behaviour changes of herring in the North Sea (Corten & van de Kamp 1991) are described in Chapter 5 section 2. Although there have been many suggestions of possible climatic effects on these changes in the North Sea fish ecosystem, it has been difficult to quantify these effects in relation to the effects of fishing and various economic and technical (marketing, gear improvements, etc.) effects.

Corten (1990) pointed out that some pelagic fish stocks in the North Sea have had long-term changes in abundance and distribution which cannot be explained by changes in fishing effort alone. He postulated that these changes could be explained by assuming a long-term reduction of Atlantic water inflow into the northern North Sea in the period 1960 to 1980 and an increase of this flow in later years. However, he could not find any physical oceanographic evidence that such a change of inflow had occurred.

Some of the long-term fluctuations in catches of pelagic fish (herring and pilchard) have been explained by long-term climatic trends in areas where the distributions of these species overlap, for example off the Devon and Cornwall coasts (Fig. 7.25). Southward *et al.* (1988) believed that the climatic change not only affected the reproduction and behaviour of the species, but also influenced their relative competitive advantages through alteration of the total ecosystem.

The North Sea herring stocks recovered rapidly in the early 1980s (Fig. 7.26), when a very low spawning stock started to produce strong year classes. Furthermore, the large stocks of groundfish (gadoids) in the 1960s showed a nearly steady decrease in 1980s, mainly owing to high rates of exploitation (Jakobsson 1991).

The semi-closed seas

The semi-closed, higher latitude, relatively shallow Baltic Sea is one of the larger water bodies which could be expected to be responsive to climatic changes, especially to global climatic changes. However, the global warming is not reflected in the long-term temperature changes in the northeast Atlantic or in the Baltic Sea. The main changes in the Baltic Sea fish ecosystem are a result of man's activities. The nearly tenfold increase of fish catches from the Baltic Sea has been discussed in Chapter 6 section 2 in connection with the eutrophication of this Sea.

Besides eutrophication, the current stagnation of deep water in the Eastern Gotland Basin since 1977 is the most pronounced observed in the Baltic Sea in this century (Matthaus 1991). It has been caused by the absence of lasting autumn storms over the Kattegat and Danish Sounds, which would force saline water into the Baltic Sea, and thus it is dependent on specific weather events rather than on climatic changes.

Catches of cod from the Baltic Sea were relatively steady in the 1960s and 1970s (about 150 000 t), peaked in the early 1980s at about 400 000 t, and from 1985 onwards there has been a steady decrease of cod catches in the southern Baltic caused by reduced recruitment and heavy fishing pressure (Jakobsson 1991). The herring fishery was steady in 1980s (about 400 000 t) and the sprat fishery showed some increase, apparently associated with reduced predation by cod.

The rapid increase of catches and biomass of sprat in the Black Sea (Fig. 7.27) and rapid decline in late 1980s (not shown in the figure) is not fully understood. The increase in the catches might have been caused by eutrophication and/or improvement and increase of the fleets, especially in Turkey, whereas decline might have been caused by the intrusion of jellyfish and comb-jellies (ctenophores) and their mass development.

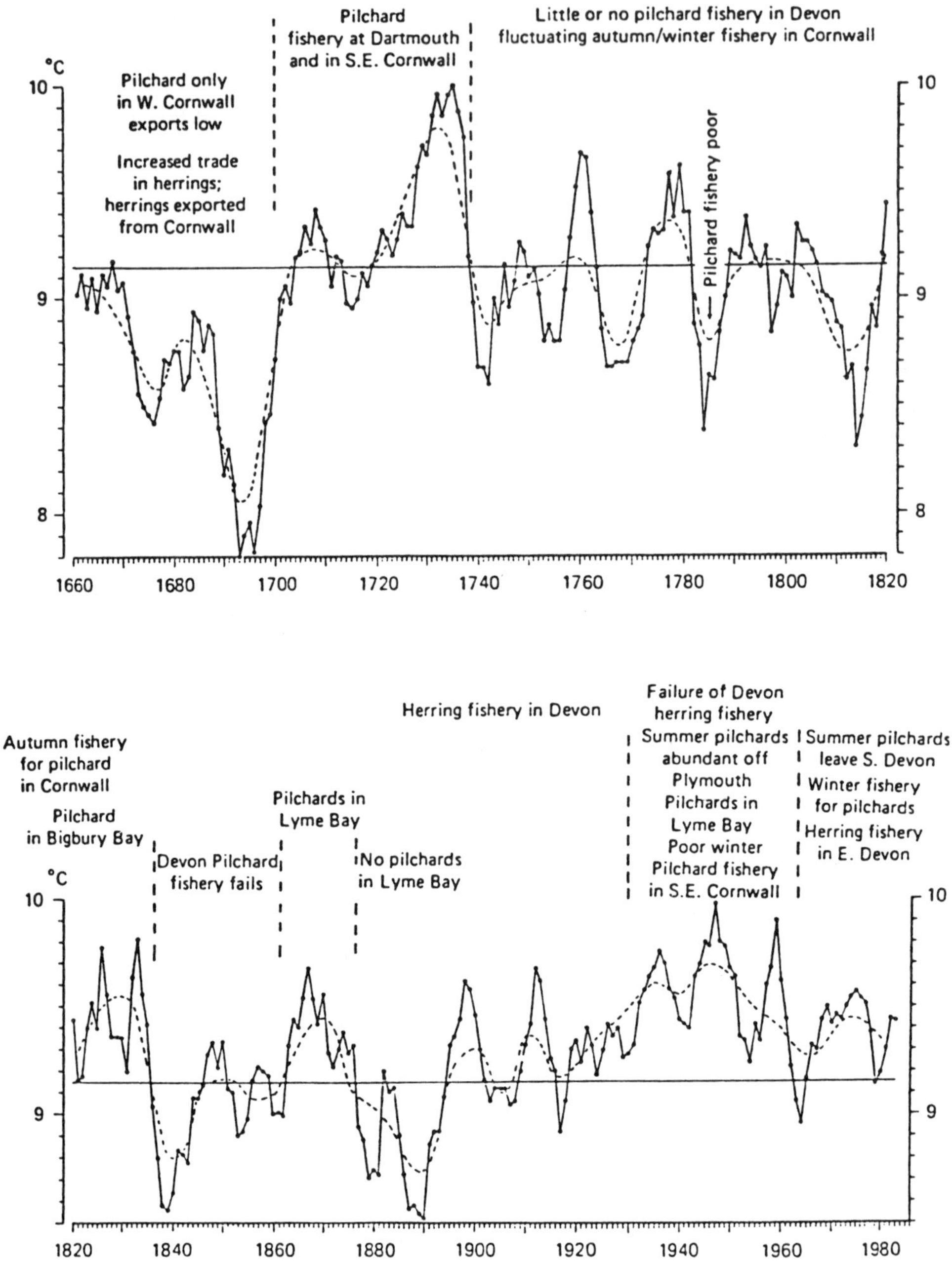

Fig. 7.25 Fishery data for 1650–1984 compared with mean air temperatures for central England. The temperatures are shown as 5-year running means (solid line) and as smoothed curves drawn through the 11-year running means (broken line). The horizontal line is the mean for the whole period (Southward *et al.* 1988).

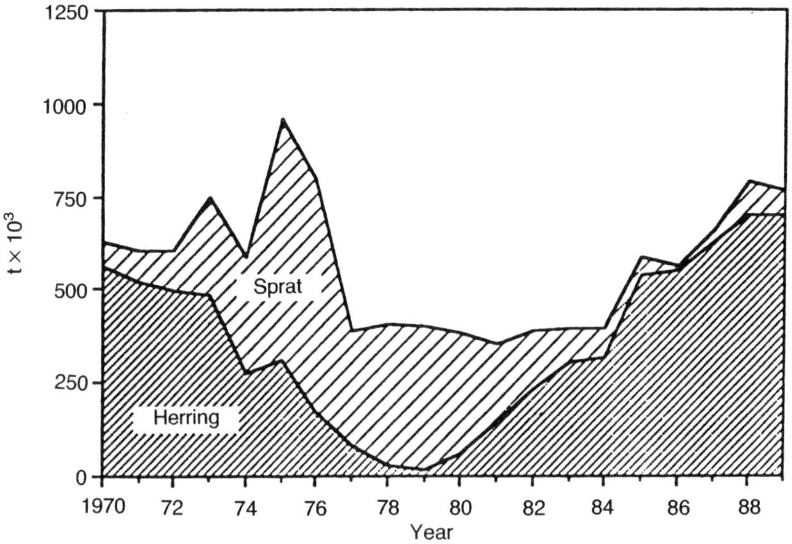

Fig. 7.26 Annual North Sea herring and sprat catches, 1970–89 (Jakobsson 1991).

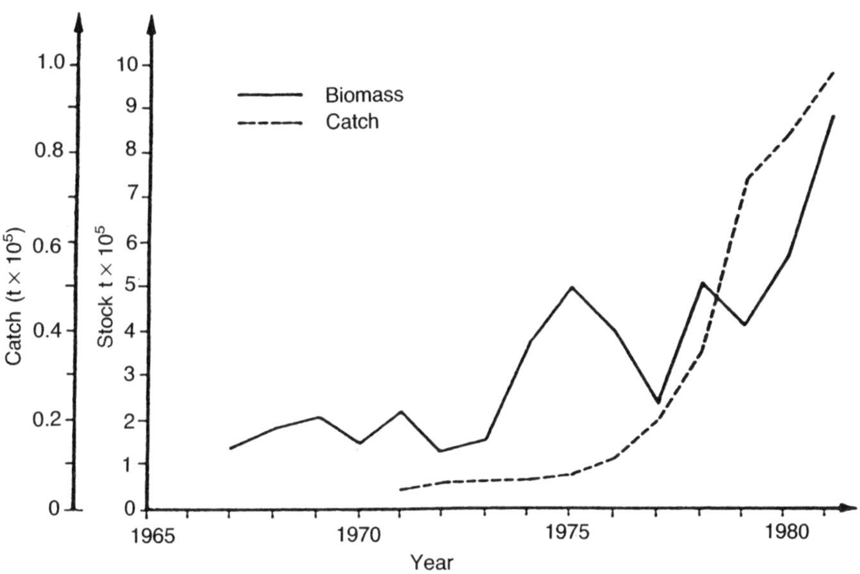

Fig. 7.27 Total catches of Black Sea sprat (1971–81) and estimates of stock biomass (1967–81) (Caddy & Griffiths 1990).

The North Pacific

The northwest Pacific has a long history of Japanese fishing and fishery research. The main fluctuations of stocks, as elsewhere in the world, are in pelagic fish (see summary by Tanaka 1983). The fluctuation of the Japanese sardine has already been described in Chapter 7 section 1. One of the earliest documented pronounced changes of pelagic fish in the northwest Pacific was that of the Hokkaido herring. This species showed large fluctuations in year-class strength and the catches were supported by few dominant year-classes which appeared every few years. The decline of this stock, which provided a peak catch of 970 000 t in 1897, started early in the twentieth century, first in southern parts of the spawning grounds, and then extended further northward. In the 1930s the frequency of occurrence and strength of stronger year classes declined markedly. However, in 1939 suddenly a large year-class came up, which dominated in the catches for 20 years. The disappearance of this year-class ended the fisheries for the Hokkaido herring. Many suggestions as to the cause of the disappearance of this stock have been made, but the factors are not yet fully understood.

The stocks of Japanese mackerel have also fluctuated (Tanaka 1983), with peak catches in the mid-1960s (1956 and 1957 year-classes), followed by a decline in the 1970s (poor year-classes from 1964 to 1968). High production from 1969 resulted in another peak, with catches of 1 000 000 t and higher in the 1970s, but catches dropped to 410 000 tonnes in 1981, possibly owing to high fishing pressure. Long-term variations in recruitment are assumed to be primarily caused by the various consequences of the meandering of the Kuroshio current.

While the mackerel catches increased, saury catches decreased. Both stocks feed on zooplankton in the Oyashio-Kuroshio frontal zone and might be competitors. The saury is a short-lived species, with a life span of at most 2 years, and therefore the total stock size depends on a single year-class and year-to-year fluctuations can be large.

The extensive fisheries developments in the northeast Pacific are very recent (from the mid-1960s) and have involved many regulations. Furthermore, the fisheries research in this area is mostly directed to satisfy prescribed management measures. Therefore, neither the catches nor the research results depict true fluctuations of stocks in this region, and neither the effects of fishing on stocks nor possible effects of climatic fluctuations can be deduced from the catch records.

Other regions and coastal fisheries

The change in catches, both amounts and species composition, from the Benguela Current region (South Africa) are shown in Figs 5.6 and 5.7. The catches increased rapidly in the 1950s and stabilized in the 1960s at about 2 000 000. The species composition changed considerably in the 1960s and 1970s. Recently conditions have been relatively stable, with anchovy providing the bulk of the catch in southern part and horse mackerel in the northern part. However, long-term alterations in the distribution of catches have been observed, attributed to changing fishing effort and in a number of instances to environmental changes. These environmental changes are related to changes in upwelling, which will cause south-north replacements of stocks (see further Chapter 5 Section 2).

The fluctuations of coastal fisheries in all regions depend on many factors, such as;

(1) population and market development,
(2) gear development,
(3) location in relation to the natural distribution boundaries of major species caught,
(4) coastal pollution, including lagoon eutrophication,
(5) fluctuations of stocks offshore, and
(6) a myriad of fishery regulations.

Therefore it is difficult to trace any climatic effects in these fisheries.

Coastal fisheries for shellfish (e.g. surf clam) are often dependent on massive settlements of larvae, which occur infrequently and irregularly (Franz 1976), the reasons for which are unclear. This applies also to larval recruitment of many finfish.

Large-scale coastal ecosystem changes can occur in upwelling regions, such as those off the Peruvian coast during the 1982–3 El Niño phenomenon (Arntz 1986).

7.3 Summary review of the past and future aspects of climate and fisheries

The recent great attention to global change and its possible consequences is in many ways misguided. There is no doubt that a slight man-made climatic change is occurring, though there is no consensus on the possible magnitude and effects of this change. Obviously some measures are needed to decrease the effects of man on the atmosphere and on Nature in general. However, surprisingly little attention has been given to the role of the ocean in climate change (Gray 1991). The change of climate and its effects are studied with atmospheric general circulation models. A recent study by a World Meteorological Organization (WMO) (1991) panel, comparing 14 such models, found that the models are giving different results and are misrepresenting, or even omitting, some mechanisms which result in their deficiency.

No global oceanic climatic changes have been found in the past 10 years which would have affected the marine fish ecosystem in the same space scale. The recent minor global warming will have no direct nor indirect noticeable effects on marine ecosystem unless it is coupled with rather profound changes in the ocean surface circulation on a global scale, which could be effected by large shifts in surface wind systems. However, considerable climatic fluctuations have occurred on a regional scale and such fluctuations undoubtedly will continue. These regional climatic fluctuations are entirely unpredictable, as weather forecasts which have any value and accuracy can be made at present only up to 10 days ahead at most.

The regional climatic fluctuations in surface weather and in the surface layers of the sea are manifested mainly in geographic shifts of the mean positions of the semi-permanent mean low systems (the Icelandic and Aleutian Lows), and in timing of the seasonal changes of surface weather and coupled processes in the surface layers of the ocean. The displacements of the positions of mean lows are also reflected in the climatic variations of the storm tracks over the oceans. The regional climatic fluctuations of the pressure and wind systems cause anomaly winds which in turn cause anomalies in ocean surface currents. The apparent results of the current anomalies are surface temperature anomalies, which are caused mainly by advection.

The causes of the long-term fluctuations in fish stocks are manifold and often complex. Some of the changes in stock abundance are influenced by hydroclime fluctuations, some are biological and ecological within the fish ecosystems, such as predation and competition, and the major cause is often man's action, fishing. The effect of fishing can be direct, i.e. lowering the biomass of the target species. However, the effect of fishing on the fish ecosystem can also be indirect, affecting the non-target species, for example by changing the dominance and competition conditions within the system and/or changing the predation conditions. The harmful effects of man-made pollution are usually local and some (e.g. eutrophication) can even be positive.

Fish stocks do not respond rapidly, either to the effects of climatic changes or to fishing, because most stocks are buffered by the presence of many age-groups in the pre-fishery juveniles and in the exploitable part of the population. For the purpose of fisheries management, most of the effects of fishing can be computed with numerical models which are not based on assumptions but composed of elements and processes where we know quantitatively the causes and effects.

A quantitative determination of the contribution of different factors to the changes in the fish ecosystem is made more difficult by the shortcomings and complexities of available data. Landing and catch data, although often used in studies of climatic effects, are affected by many factors other than fish abundance and availability, such as market requirements, technological developments, and fisheries regulations. A great need exists to know the properties and limitations of all fisheries data and to weigh and analyse those data carefully, bearing in mind the known limitations.

Large fluctuations in abundance have occurred world-wide in pelagic (neritic) fish stocks. These fluctuations are not synchronous or global but regional, as are their causes. A common characteristic of these fluctuations is the replacement of one dominant species by an ecologically similar species, such as sardines versus anchovies. Further characteristics are the change in spatial distribution and, in some cases, even changes in annual and life-cycle migration routes. Heavy fishing pressure is assumed in most cases to be chiefly responsible for the collapse of most pelagic stocks. However, other causes cannot be ruled out, especially where large recruitment fluctuations occur in short-lived species. In earlier decades, it was often assumed that larval recruitment fluctuations affecting the stock abundance might have been caused mainly through availability of proper food for larvae. This cause is now considered less likely as mariculture experiments have shown that high survival of larvae of many species is possible with relatively low concentrations of proper food. Furthermore, our knowledge of the abundance distribution of pelagic (planktonic) food in space and time is incomplete. The availability of pelagic food has, however, contributed to fluctuations of pelagic stocks in upwelling regions (e.g. El Niño) and possibly of the Atlanto-Scandian herring. Most of the causes of the fluctuations of neritic stocks are, however, still obscure.

There are a few cases of changes of fish stocks where regional climatic fluctuations seem to be the main cause of stock fluctuation (e.g. West Greenland cod), and there are a few other cases where climatic changes of wind system influence, via surface current (advectional) changes, recruitment of some species (e.g. Barents Sea cod). However, climatic fluctuation alone might not be the direct cause of the profound changes in the carrying capacity of the Barents Sea in the late 1980s. The carrying capacity of an area in respect of the fish ecosystem is determined by the production and availability not only of basic food (plankton and benthos), but also of forage fish. The decline of the latter might be the main cause of the quasi-catastrophic event in the Barents Sea in the 1980s.

Changes in fish ecosystem cannot be easily predicted. Even man's effects (fishing) are difficult to predict as well as detect, as fishing affects not only target species but other species as well. There are also natural fluctuations in systems (e.g. caused by competition and predation) where man cannot interfere.

Better directed research on behaviour and interactions in the fish ecosystem is needed. Recent research has shown that old truths must also be relearned. For example, recent mariculture research has shown that maturation and growth of some fish might be directly affected by light, and so the amount of sunshine might be one of the environmental factors influencing the fish ecosystem, one not seriously considered in the past studies. It must be realized that climatic changes and fluctuations can be favourable as well as unfavourable to the marine ecosystem from the human point of view. Furthermore, the marine fish ecosystem is not stable, but fluctuates considerably even without noticeable environmental changes or effects of fishing.

Larval recruitment might be the life stage of fish most sensitive to environmental changes. This recruitment can be affected by a multitude of factors in space and time, and it is usually not possible to pinpoint the factor or stage or location where and when the environment has or has not had the greatest effect on survival. The juveniles of fish between the larval stage and the age/size at which the species becomes fully recruited to the fishery are subjected to environmental effects, but are more subject to predation which might often be the main cause of juvenile mortality. Predation is also partly affected by the environment and its changes.

Single species fisheries management has not been fully successful in the past for a number of reasons, such as failure to consider species to species interactions, especially predation, and the exclusion of the effects of environment and its anomalies, including regional climatic fluctuations, on the fluctuations of fish stocks. Thus we must consider all the factors and processes which affect the fish ecosystem, segregate them by cause and effect and construct quantitative simulation models which will help us to manage the system to some extent.

Total catches of fish cannot be increased by noticeable amounts in most ocean regions in the world. However, there will be fluctuations in stock sizes of most species, caused by several factors, which we must learn to recognize and quantify. Furthermore, in order to obtain the maximum amount out of a given regional fish ecosystem, we must build in flexibility in our fisheries so that targeting can be switched easily from one species to another, possibly in another area. This would also require considerably more flexibility in fisheries management.

One of the main tasks of fisheries is to learn to manipulate the marine fish ecosystem so as to obtain the maximum return from it to man and still maintain its integrity, just as we manipulate the terrestrial ecosystem.

References

Note

Secondary sources (e.g. for information within figures reproduced in the text) are not included in the reference list; for full details see the primary sources listed.

Alverson, D.L. (1960) A study of annual and seasonal bathymetric catch patterns for commercial important groundfishes of the Pacific northwest coast of North America. *Pacific Mar. Fish. Comm. Bull.*, **4**, 1–66.

Alverson, D.L. (1991) Commercial fisheries and the Steller Sea Lion (*Eumetopias jubatus*): The conflict arena. Fisheries Research Institute, Univ. of WA. School of Fisheries, FRI-UW-9106, 90 pp.

Andersen, K.P. & Ursin, E. (1977) A multispecies extension to the Beverton and Holt theory of fishing, with account of phosphorus circulation and primary production. *Medd. Danm. Fisk.-og Havunders.*, N.S. **7**, 319–435.

Anon. (1969) Report on the joint investigations by Iceland, Norway and USSR on the distribution of herring in relation to hydrography and plankton in the Norwegian Sea May/June 1969.

Anon. (1991) Tapte redskap fisker mye. *Fiskets Gang*, **77**(11), 33–5.

Arntz, W.E. (1986) The two faces of El Niño 1982–83. *Meeresforschung – Reports on Marine Research*, **1**, 1–46.

Arntz, W.E. & Valdivia, J. (1985) Vision integral del problema – El Niño –: Introduccion 'El Niño' Su Impacto en la Fauna Marina. *Bul. Inst. Mar. Peru*, Spec. Vol. 5–9.

Bagge, O. (1981) The yearly consumption of cod in the Baltic and the Kattegat as estimated from stomach content. ICES, C.M. 1981/J:27, 31 pp.

Bakken, E. (1983) Recent history of Atlanto-Scandian herring stocks. FAO Fisheries Report, **291**(2), 521–36.

Balls, R. (1960) Wind and change over East Anglia? *World Fishing*, (10).

Bax, N.J. & Laevastu, T. (1990) Biomass potential of large marine ecosystems. A systems approach. In *Large Marine Ecosystems* (Ed. by K. Sherman, L.M. Alexander & B.D. Gold), pp. 188–205. AAAS.

Bell, F. & Pruter, A. (1958) Climatic temperature changes and commercial yields of some marine fisheries. *J. Fish. Res. Bd Canada*. **15**(4), 625–683.

Belveze, H. & Erzini, K. (1983) The influence of hydroclimatic factors on the availability of the sardine. (*Sardina pilchardus Walbaum*) in the Moroccan Atlantic fishery. *FAO Fisheries Report*, **291**(2), 285–328.

Bengtsson, L. (1981) The weather forecast. *Review Pure and Applied Geophysics*, **119**(3), 515–37.

Bergstad, O.A., Jörgenson, T. & Dragesund, O. (1985) Life history features and ecology of the gadoid resources of the Barents Sea. Gadoid Workshop, Seattle, 1985. Mimeo.

Bernard, R.D. (1973) Porpoises predict storms. *Underwat. Nat.*, **8**(1), 43–4.

Berry, F.A., Haggard, W.H. & Wolff P.M. (1954) Description of contour patterns at 500 millibars. Bureau of Aeronautics Project AROWA (TED-UNL-MA-501). AFSWP Report 717, 116 pp.

Beverton, R.J.H. & Lee, A.J. (1965) The influence of hydrographic and other factors on the distribution of cod on the Spitsbergen Shelf. *ICNAF Spec. Publ.*, **6**, 225–45.

Binet, D. & Orstom, A. (1986) French sardine and herring fisheries: A tentative description of their fluctuations since the eighteenth century. *Int. Symp. Long Term Changes Mar. Fish. Pop.*, 253–71. Vigo 1986.

Blindheim, J. (1967) Hydrographic fluctuations off West Greenland during the years 1959–1966. ICNAF Res. Doc. 67/70, Serial No. 1814, 4 pp.

Blindheim, J., Loeng, H. & Saetre, R. (1981) Long-term temperature trends in Norwegian coastal waters. ICES, C.M. 1981/C:19, 13 pp.

Bocharov, L.N. (1990) *Systems Analysis in Short-term Fishery Forecasts* (in Russian). Nauka, St Petersburg, 208 pp.

Brown, B.E., Anthony, V.C., Anderson, E.D., Hennemuth, R.C. & Sherman, K. (1983) The dynamics of pelagic fishery resources off the northwestern coast of the United States under conditions of extreme fishing perturbations. *FAO Fisheries Report*, **291**(2), 465–506.

Buch, E. & Hansen, H. (1986) Climate and cod fishery at West Greenland. Mimeo 14 pp.

Burbank, A. & Douglass, R.H. (1969) Fisheries Forecasting Systems – A study of the Japanese fisheries forecasting system. Final report for IR & D MJO 9843-39. TRW Systems, 99900–6865–RO–00, 55 pp.

Burd, A.C. (1978) Long-term changes in North Sea herring stocks. *Rapp. P.-V. Réun. Cons. Int. Explor. Mer*, **172**, 137–53.

Burgner, R.L. (1980) Some features of ocean migrations and timing of Pacific salmon. In *Salmonid Ecosystems of the North Pacific* (Ed. by W. McNeil & C. Himsworth), pp. 153–164. Oregon State Univ. Press, Cervallis.

Cabelli, V.J., Dufour, A.P., Levin, M.A. & Habermann, P.W. (1976) The impact of pollution on marine bathing beaches: An epidemiological study. *Am. Soc. Limnol. Oceanogr., Spec. Symp.* **2**, 424–32.

Caddy, J.F. & Griffiths, R.C. (1990) Recent trends in the fisheries and environment in the General Fisheries Council for the Mediterranean (GFCM) area. *Studies and Reviews General Fisheries Council for the Mediterranean*, No. 63. FAO, Rome, 71 pp.

Carruthers, J.N. (1938) Fluctuation in the herring of the East Anglian autumn fishery, the yield of the Ostend spent herring fishery, and the haddock of the North Sea — in the light of relevant wind conditions. *Rapp. P.-V. Réun. Cons. Perm. Int. Explor. Mer*, **107**(3), 10–15.

Carruthers, J.N. (1951) An attitude on 'fishery hydrography'. *J. Mar. Res.*, **10**(1), 101–18.

Carruthers, J.N. (1966) Science and the fisherman. *Fishing News International*, Jan.–Mar. 1966, 6 pp.

Chapman, W.M. (1962) Oceanography and fishing. *Navigation*, **9**(3), 200–4.

Chase, J. (1955) Winds and temperatures in relation to the brood-strength of Georges Bank haddock. *J. Cons. Int. Explor. Mer*, **21**, 17–24.

Christensen, V. (1983) Predation by sand eel on herring larvae. ICES, C.M. 1983/L:27, 10 pp.

Clark, R.B. (1986) *Marine Pollution*. Clarendon Press, Oxford, 215 pp.

Colebrook, J.M. (1979) Continuous plankton records: seasonal cycles of phytoplankton and copepods in the North Atlantic Ocean and North Sea. *Marine Biology*, **51**, 23–32.

Colijn, F. (1991) Changes in plankton communities: When, Where and Why? *ICES, Variability Symp.*, No. 13, 20 pp.

Corten, A. (1983) Predation on herring larvae by the copepod *Candacia armata*. ICES, C.M. 1983/H:20, 3 pp.

Corten, A. (1990) Long-term trends in pelagic fish stocks of the North Sea and adjacent waters and their possible connection to hydrographic changes. *Netherlands J. Sea Res.*, **25**, 227–35.

Corten, A. & Kamp, G. van de (1991) Natural changes in pelagic fish stocks of the North Sea in the 1980s. *ICES, Variability Symp.*, No. 27, 11 pp.

CPR Survey Tem. (1991) Continuous plankton records: the North Sea in the 1980s. *ICES, Variability Symp.*, No. 17, 5 pp.

Crawford, R.J.M., Shelton, P.A. & Hutchings, L. (1983) Aspects of variability of some neritic stocks in the southern Benguela system. *FAO Fisheries Report*, **291**(2), 407–48.

Crawford, R.J.M., Shannon, L.V. & Pollock, D.E. (1987) The Benguela ecosystem. Part IV, the major fish and invertebrate resources. *Oceanogr. Mar. Biol. Ann. Rev.*, **25**, 353–505.

Cushing, D.H. (1980) The decline of the herring stocks and the gadoid outburst. *J. Cons. Int. Explor. Mer.*, **39**(1), 70–81.

Cushing, D.H. (1982a) *Climate and Fisheries*. Academic Press, London, 373 pp.

Cushing, D.H. (1982b) A simulacrum of the Iceland cod stock. *J. Cons. Int. Explor. Mer.*, **40**(1), 27–36.

Cushing, D.H. (1986) The Northwest Atlantic cod fishery. *Int. Symp. Long Term Changes Mar. Fish Pop.*, pp 83–94. Vigo 1986.

Cushing, D.H. (1988) The northerly wind. In *Toward a Theory on Biological-Physical Interactions in the World Ocean* (Ed. by B.J. Rothschild), pp. 235–244. Kluwer Academic Publishers, Dordrecht.

Cushing, D.H. & Dickson, R.R. (1976) The biological response in the sea to climatic changes. *Adv. Mar. Biol.*, **14**, 1–122.

Daan, N. (1976) Some preliminary investigations into predation of fish eggs and larvae in the southern North Sea. ICES, C.M. 1976/L:15, 11 pp.

Daan, N. (1980) A review of replacement of depleted stocks by other species and the mechanisms underlying such replacement. *Rapp. P.-V. Réun. Cons. Int. Explor. Mer.*, **177**, 405–21.

Daan, N. (1981) Feeding of North Sea cod in roundfish area 6 in 1980 – Preliminary results. ICES, C.M. 1981/G:73, 4 pp.

Daan, N., Rijnsdorp, A.D. & Overbeeke, G.R. van. (1985) Predation by North Sea herring *Clupea harengus* on eggs of plaice *Pleuronectes platessa* and cod *Gadus morhua*. *Trans. Am. Fish. Soc.*, **114**, 499–506.

Dethlefsen, V. & Tiews, K. (1985) Review on the effects of pollution on marine fish life and fisheries in the North Sea. *Zeitschr. angew. Ichthyol.*, **3**(1985), 97–118.

Dickson, R.R. (1968) Long-term changes in hydrographic conditions in the North Atlantic and adjacent seas. ICES, C.M. 1968/C:27, 7 pp.

Dickson, R.R. (1971) A recurrent and persistent pressure-anomaly pattern as the principal cause of intermediate-scale hydrographic variation in the European shelf seas. *Dtsch. Hydrogr. Zeitschr.*, **24**(3), 97–119.

Dickson, R.R. (1991) Recent changes in the summer plankton of the North Sea. *ICES, Variability Symp.*, No. 16, 25 pp.

Dickson, R.R. & Lamb, H.H. (1972) A review of recent hydrometeorological events in the North Atlantic sector. *ICNAF Spec. Publ.*, **8**, 35–62.

Dickson, R.R. & Lee, A. (1969) Atmospheric and marine climate fluctuations in the North Atlantic region. Progress in Oceanography, (5), (Ed. Mary Sears) Pergamon Press, 55–65.

Dickson, R.R. & Namias, J. (1976) North American influences on the circulation and climate of the North Atlantic sector. *Monthly Weather Review*, **104**(10), 1255–65.

Dietrich, G. & Kalle, K. (1957) *Allgemeine Meereskunde*. Borndraeger, Berlin-Nikolassee, 492 pp.

Dietrich, G., Sahrhage, D. & Schubert, K. (1957) The localization of fish concentrations by thermometric methods. *International Fishing Gear Congress 1957*, 7–12 Oct., Hamburg, Germany. Paper No. 34-(b), 20 pp.

Dow, R.L. (1964) A comparison among selected marine species of an association between sea water temperature and relative abundance. *J. Cons. Int. Explor. Mer.*, **28**, 425–31.

Dragesund, O., Hamre, J. & Ulltang, Ø. (1980) Biology and population dynamics of the Norwegian spring-spawning herring. *Rapp. P.-V. Réun. Cons. Int. Explor. Mer.*, **177**, 43–71.

Elizarov, A.A. (1980) Oceanographic prerequisites for seasonal variability prediction of the biological and fishing productivity. In *Moscow, 'Pishchevaya Promyshlennost'*, 118 pp. Translated by ESDUCK, Cairo, 1981, pp. 69–99.

Ellett, D.J. & Blindheim, J. (1991) Climate and hydrographic variability in the ICES area during the 1980s. *ICES, Variability Symp.*, No. 1, 12 pp.

Elmgren, R. (1984) Trophic dynamics in the enclosed, brackish Baltic Sea. *Rapp. P.-V. Réun. Cons. Int. Explor. Mer.*, **183**, 152–69.

Elmgren, R. (1989) Man's impact on the ecosystem of the Baltic Sea: Energy flows today and at the turn of the century. *Ambio*, **18**(6), 326–32.

Favorite, F. & Ingraham, W.J. Jr. (1976) Sunspot activity and oceanic conditions in the northern North Pacific Ocean. *J. Oceanogr. Soc. Japan*, **32**(3), 107–15.

Favorite, F. & Ingraham, W.J. Jr. (1978) Sunspot activity and oceanic conditions in the northern North Pacific Ocean. In *Ocean Variability: Effects on U.S. Marine Fishery Resources, 1975* (Ed. by J.R. Goulet & E.D. Haynes). *NOAA Tech. Rep. NMFS Circ.*, **416**, 191–6.

Favorite, F. & Laevastu, T. (1981) Finfish and the environment. *The Eastern Bering Sea Shelf: Oceanography and Resources* (Ed. by Donald W. Hood & John A. Calder). *NOAA*, **1**, 597–610.

Flintegaard, H. (1981) An estimate of the food consumption of whiting (*Merlangius merlangus*). ICES, C.M. 1981/G:81, 11 pp.

Flohn, H. (1965) Research aspects of long-range forecasting. *WMO Tech. Note*, No. 66.

Flohn, H. & Fantechi, R. (Eds) (1984) *The Climate of Europe: Past, Present and Future*. D. Reidel Publishing Company, Dordrecht, 356 pp.

Franz, D.R. (1976) Distribution and abundance of inshore populations of the surf clam *Spisula solidissima*. *Am. Soc. Limnol. Oceanogr., Spec. Symp.* **2**, 404–413.

Fraser, J.H. (1961) The oceanic and bathypelagic plankton of the north-east Atlantic and its possible significance to Fisheries. *Mar. Res.*, 4, 48 pp.

Furness, R.M. (1989) Estimation of fish consumption by seabirds and the influence of changes in fish stock size on seabird predation rates. *ICES, 1989 MSM Symp.*, No. 36, 8 pp.

Gammelsrød, T., Osterhus, S. & Godoy, O. (1991) Forty years of hydrographic observations at weather ship station *Mike* (66°N., 2°E.). *ICES, Variability Symp.*, Poster No. 35.

George, J.J. & Wolff, P.M. (1955) The prediction of cyclone intensity over the North Atlantic. Bur. of Aeronautics Project AROWA (TED-UNL-MA-501), Research Report on Task 13, 52 pp.

Glover, R.S., Robinson, G.A. & Colebrook, J.M. (1971) Plankton in the North Atlantic — An example of the problems of analysing variability in the environment. *NATO Science Committee Conference, North Sea Science Working Papers*, **1**, 14 pp.

Glover, R.S., Robinson, G.A. & Colebrook, J.M. (1974) Marine biological surveillance. *Environment and Change* Feb., 395–402.

Grainger, R.J.R. (1978) Herring abundance off the west of Ireland in relation to oceanographic variation. *J. Cons. Int. Explor. Mer.*, **38**(2), 180–8.

Gray, J.S. (1982) Effects of pollutants on marine ecosystems. *Netherl. J. Sea Res.*, **16**, 424–43.

Gray, J.S. (1991) Climate change. *Marine Poll. Bull.*, **22**(4), 169–71.

Gray, J.S., Clarke, K.R., Warwick, R.M. & Hobbs, G. (1990) Detection of initial effects of pollution on marine benthos: an example from the Ekofisk and Eldfish oilfields, North Sea. *Mar. Ecol. Prog. Ser.*, **66**, 285–99.

Greer Walker, M., Harden Jones, F.R. & Arnold, G.P. (1978) The movement of plaice (*Pleuronectes platessa* L.) tracked in the open sea. *J. Cons. Int. Explor. Mer.*, **38**(1), 58–86.

Hamre, J. (1988) Some aspects of the interrelation between the herring in the Norwegian Sea and the stocks of capelin and cod in the Barents Sea. ICES, C.M. 1988/H:42, 15 pp.

Harden Jones, F.R. & Scholes, P. (1980) Wind and the catch of a Lowestoft trawler. *J. Cons. Int. Explor. Mer.*, **39**(1), 53–69.

Harden Jones, F.R., Scholes, P. & Cheeseman, C. (1969) An apparent effect of wind on the catch of a Lowestoft trawler. ICES, C.M. 1969/B:14, 3 pp.

Harden Jones, F.R., Greer Walker, M. & Arnold, G.P. (1976) Tactics of fish movement in relation to migration strategy and water circulation. In *Advances in Oceanography, Proceedings of Oceanographic Assembly 1976* (Ed. by H. Charnock & Sir George Deacon), pp. 185–207. Blenheim Press, New York.

Hasse, L. (1974) On the surface to geostrophic wind relationship at sea and the stability dependence of the Resistance Law. *Beitrage zur Physik der Atmosphäre*, **47**, 45–55.

Helland-Hansen, B. & Nansen, F. (1920) Temperature variations in the North Atlantic Ocean and in the atmosphere. Introductory Studies on the Causes of Climatological Variations. *Smith. Misc. Coll.*, **70**(4), 1–408.

Hempel, G. (1973) Food requirements for fish production. In *Marine Food Chains* (Ed. by J.H. Steele). Oliver & Boyd, Edinburgh, pp. 255–60.

Hempel, G. (1978a) North Sea fisheries and fish stocks — A review of recent changes. *Rapp. P.-V. Réun. Cons. Int. Explor. Mer.*, **173**, 145–67.

Hempel, G. (1978b) Synopsis of the symposium on North Sea fish stocks — recent changes and their causes. *Rapp. P.-V. Réun. Cons. Int. Explor. Mer.*, **172**, 445–9.

Herman, F. (1967) Temperature variations in the West Greenland area after 1950. ICNAF Res. Doc. 67/59, Serial No. 1849, 12 pp.

Hicks, S.D. (1973) Trends and variability of yearly mean sea level 1893–1971. NOAA Tech. Memo NOS 12, 13 pp.

Hodgson, W.C. (1957) *The Herring and its Fishery*. Routledge & Kegan Paul, London, 197 pp.

Hohn, R. (1971) On the climatology of the North Sea. *NATO Science Committee Conference, North Sea Science Working Papers*, Vol. 1, 16 pp.

Holden, M.J. (1978) Long-term changes in landings of fish from the North Sea. *Rapp. P.-V. Réun. Cons. Int. Explor. Mer.*, **172**, 11–26.

Hubert, W.E. & Laevastu, T. (1970) Synoptic analysis and forecasting of surface currents. U.S. Navy Weather Research Facility 36–0667–127, 47 pp.

ICES (1984) Report of the Working Group on Cod Stocks off East Greenland. ICES, C.M. 1984/Assess: 5, 25 pp.

ICES (1989) Report of the Multispecies Assessment Working Group. ICES, C.M. 1989/ Assess:20, pp. 126–176.

Jakobsson, J. (1979) The North Icelandic herring fishery and environmental conditions 1960–1968. *ICES, Symp. Biol. Basis of Pelagic Fish Stock Management*, No. 30, 101 pp.

Jakobsson, J. (1980) Exploitation of the Icelandic spring- and summer-spawning herring in relation to fisheries management, 1947–1977. *Rapp. P.-V. Réun. Cons. Int. Explor. Mer.*, **177**, 23–42.

Jakobsson, J. (1991) Recent variability in fisheries of the North Atlantic. *ICES, Variability Symp.*, No. 21, 55 pp.

Johannessen, A. (1980) Predation on herring (*Clupea harengus*) eggs and young larvae. ICES, C.M. 1980/H:33, 12 pp.

Johnston, R. (1977) What North Sea oil might cost fisheries. *Rapp. P.-V. Réun. Cons. Int. Explor. Mer.*, **171**, 212–23.

Jones, R. & Hislop, J.R.G. (1978) Changes in North Sea haddock and whiting. *Rapp. P.-V. Réun. Cons. Int. Explor. Mer.*, **172**, 58–71.

Josefson, A.B. (1990) Increase of benthic biomass in the Skagerrak-Kattegat during the 1970s and 1980s — effects of organic enrichment? *Mar. Ecol. Prog. Ser.*, **66**, 117–30.

Kahma, K. & Voipio, A. (1989) Seasonal variation of some nutrients in the Baltic Sea and the interpretation of monitoring results. ICES, C.M. 1989/C:31, 8 pp.

Kalejs, M. & Ojaveer, E. (1989) Long-term fluctuations in environmental conditions and fish stocks in the Baltic. *Rapp. P.-V. Réun. Cons. Int. Explor. Mer.*, **177**, 332–54.

Klimaj, A. (1976) *Fishery Atlas of the Northwest African Shelf*. Foreign Scientific Publications Dept of the National Center for Scientific, Technical and Economic Information, Warsaw, 221 pp.

Knauss, J.A. (1960) Observations of irregular motion in the open ocean. *Deep-Sea Res.* 7(1), 68–69.

Kondo, K. (1980) The recovery of the Japanese sardine — The biological basis of stock-size fluctuations. *Rapp. P.-V. Réun. Cons. Int. Explor. Mer.*, **177**, 332–54.

Koslow, J., Thompson, K.R. & Silvert, W. (1987) Recruitment to Northwest Atlantic cod (*Gadus morhua*) and haddock (*Melanogrammus aeglefinus*) stocks: Influence of stock size and climate. *Can. J. Fish. Aquat. Sci.*, **44**, 26–39.

Laevastu, T. (1960) Factors affecting the temperature of the surface layer of the sea. *Soc. Sci. Fennica, Comment. Physico-Matem.* **25**(1), 1–136.

Laevastu, T. (1961) Weather and fisheries. *WMO Bulletin*, July 1961, 144–50.

Laevastu, T. (1983) Numerical simulation in fisheries oceanography with reference to the Northeast Pacific and the Bering Sea. In *From Year to Year* (Ed. by Warren S. Wooster). Washington Sea Grant Publication, 180–195.

Laevastu, T. (1984) The effects of temperature anomalies on the fluctuation of stocks. *Rapp. P.-V. Réun. Cons. Int. Explor. Mer.*, **185**, 214–25.

Laevastu, T. & Bax, N. (1989) Predation controlled recruitment in Bering Sea fish ecosystem: Numerical simulation and empirical evidence. *ICES, MSM Symp.*, No. 8, 24 pp.

Laevastu, T. & Favorite, F. (1976) Dynamics of pollock and herring biomasses in the eastern

Bering Sea. NW and Alaska Fish. Center, Seattle, Proc. Rpt, 50 pp.
Laevastu, T. & Favorite, F. (1988) *Fishing and Stock Fluctuations*. Fishing News Books, Oxford, 239 pp.
Laevastu, T. & Hayes, M.L. (1981) *Fisheries Oceanography and Ecology*. Fishing News Books Oxford, 199 pp.
Laevastu, T. & Hela, I. (1970) *Fisheries Oceanography*. Fishing News Books Oxford, 238 pp.
Laevastu, T. & Hubert, W.E. (1970) The nature of sea surface anomalies and their possible effects on weather. Fleet Numerical Weather Central, Tech. Note No. 55, 17 pp.
Laevastu, T. & Marasco, R. (1985) Evaluation of the effects of oil development on the commercial fisheries in the eastern Bering Sea. NW and Alaska Fish. Centre, Seattle, Proc. Rpt 85-19, 40 pp.
Laevastu, T. & Marasco, R. (1992) Trophic interactions of marine mammals and birds with fishery resources in the Eastern Bering Sea. (MS report in press), 18 pp.
Laevastu, T., Ingraham, J. & Favorite, F. (1986) Surface wind anomalies and their possible effects on fluctuation of fish stocks via recruitment variations. *Int. Symp. Long Term Changes Mar. Fish Pop.*, pp. 393–414. Vigo, 1986.
Lamb, H.H. (1984a) Some studies of the little ice age of recent centuries and its great storms. In *Climatic Changes on a Yearly to Millenial Basis* (Ed. by N.A. Mörner & W. Karlen), pp. 309–29. D. Reidel Publishing Co., Dordrecht.
Lamb, H.H. (1984b) Climate and history in northern Europe and elsewhere. In *Climatic Changes on a Yearly to Millenial Basis* (Ed. by N.A. Mörner & W. Karlen), pp. 225–40. D. Reidel Publishing Co., Dordrecht.
Lamb, H.H. & Ratcliffe, R.A.S. (1969) On the magnitude of climatic anomalies in the oceans and some related observations of atmospheric circulation behaviour. ICES, C.M. 1969, No. 19, 19 pp.
Lapointe, M.F. & Peterman, R.M. (1991) Spurious correlations between fish recruitment and environmental factors due to errors in the natural mortality rate used in virtual population analysis (VPA). *ICES J. Mar. Sci.*, **48**(2), 219–28.
Larkin, P.A. (1977) An epitaph for the concept of maximum sustained yield. *Trans. Am. Fish. Soc.*, **106**,(1), 1–10.
Larson, S. & Laevastu, T. (1976) Monthly mean evaporation and surface winds over the Northern Hemisphere oceans and their year-to-year variations. WMO, ISBN 92–63–10442-5, pp. 66–95.
Larsson, U., Elmgren, R. & Wulff, F. (1985) Eutrophication and the Baltic Sea: Causes and Consequences. *Ambio* (1985) **14**(1), 10–14.
Lasker, R. (1981) Factors contributing to variable recruitment of the northern anchovy (*Engraulis mordax*) in the California Current: contrasting years 1975 through 1978. *Rapp. P.-V. Réun. Cons. Int. Explor. Mer.*, **178**, 375–88.
Laurs, M.R. & Lynn, R.J. (1977) Seasonal migration of North Pacific albacore, *Thunnus alalunga*, into North American coastal waters: Distribution, relative abundance, and association with transition zone waters. *Fish. Bull.*, **75**(4), 795–822.
Laurs, M.R., Yuen, H.S.H. & Johnson, J.H. (1977) Small-scale movements of albacore, *Thunnus alalunga*, in relation to ocean features as indicated by ultrasonic tracking and oceanographic sampling. *Fish. Bull.*, **75**(2), 347–55.
LeClus, F. (1990) Impact and implications of large-scale environmental anomalies on the spatial distribution of spawning of the Namibian pilchard and anchovy populations. *S. African J. Mar. Sci.*, **9**, 141–59.
LeClus, F. (1991) Hydrographic features related to pilchard and anchovy spawning in the northern Benguela system, comparing three environmental regimes. *S. African J. Mar. Sci.*, **10**, 103–24.
Lee, A. (1978) Effects of man on the fish resources of the North Sea. *Rapp. P.-V. Réun. Cons. Int. Explor. Mer.*, **173**, 231–40.
Leggett, W.C., Frank, K.T. & Carscadden, J.E. (1984) Meteorological and hydrographic regulation of year-class strength in capelin (*Mallotus villosus*). *Can. J. Fish. Aquat. Sci.*, **41**, 1193–201.

Lett, P.F., Kohler, A.C. & Fitzgerald, D.N. (1975) Role of stock biomass and temperature in recruitment of Southern Gulf of St Lawrence Atlantic cod, *Gadus morhua*. *J. Fish. Res. Bd Can.*, **32**, 1613–27.

Lindquist, A. (1983) Herring and sprat: Fishery independent variations in abundance. *FAO Fisheries Report*, **291**(3), 813–22.

Lluch-Belda, D., Crawford, R.J.M., Kawasaki, T., MacCall, A.D., Parrish, R.H., Schwartziose, R.A. & Smith, P.E. (1989) World-wide fluctuations of sardine and anchovy stocks: The regime problem. *S. Afr. J. Mar. Sci.*, **8**, 195–205.

Loeng, H. (1989) The influence of temperature on some fish population parameters in the Barents Sea. *J. Northwest Atl. Fish. Sci.*, **9**, 103–13.

Loeng, H., Nakken, O. & Raknes, A. (1983) The distribution of capelin in the Barents Sea in relation to the water temperature in the period 1974–1982. *Fisken Hav.*, 1983(1), 1–17.

Loeng, H., Blindheim, J., Adlandsvik, B. & Otterson, G. (1991) Climatic variability in the Norwegian and Barents Seas. *ICES, Variability Symp.*, No. 4, 19 pp.

Lough, R.G., Bolz, G.R., Grosslein, M.D. & Potter, D.C. (1981) Abundance and survival of sea herring (*Clupea harengus*) larvae in relation to environmental factors, spawning stock size, and recruitment for the Georges Bank area, 1968–1977 seasons. *Rapp. P.-V. Réun. Cons. Int. Explor. Mer.*, **178**, 220–2.

Lumb, F.H. (1963) A simple method of estimating wave height and direction over the North Atlantic. *Mar. Obs.*, Jan. 1963, 23–9.

Lundbeck, J. (1962) Biologish-statistische Untersuchungen über die Deutsche Hochseefischerei. IV, 5: Die Dampferfischerei in der Nordsee. *Ber. Dr. Wiss. Komm. Meeresforsch.*, **16**, 177–246.

MacCall, D. (1983) Variability of pelagic fish stocks off California. *FAO Fisheries Report*, **291**(2), 101–12.

McLain, D.R. & Favorite, F. (1976) Anomalously cold winters in the southeastern Bering Sea 1971–1975. *Marine Science Communications*, **2**(5), 299–334.

MacPherson, E., Maso, M., Barange, M. & Gordoa, A. (1992) Relationship between measurements of hake biomass and sea surface temperature of Southern Namibia. *S. Afr. J. Mar. Sci.* (in press).

Magnusson, K.G. & Pálsson, O.K. (1989) Trophic ecological relationships of Icelandic cod. *Rapp. P.-V. Réun. Cons. Int. Explor. Mer.*, **188**, 206–24.

Makkonen, L., Launiainen, J., Kahma, K. & Alenius, P. (1984) Long-term variations in some physical parameters of the Baltic Sea. In *Climatic Changes on a Yearly to Millenial Basis* (Ed. by N.A. Mörner & W. Karlen), pp. 391–399. D. Reidel Pub. Co., Dordrecht.

Mälkki, P. & Tamsalu, R. (1985) Physical features of the Baltic Sea. *Finnish Marine Research*, **252**, 110 pp.

Malmberg, S.Aa. (1969) Hydrographic changes in the waters between Iceland and Jan Mayen in the last decade. *Jokull*, **19**, 30–43.

Malmberg, S.Aa. & Svansson, A. (1982) Variations in the physical marine environment in relation to climate. *ICES*, C.M. 1982/Gen:4, 44 pp.

Malmberg, S.Aa. & Kristmannsson, S.S. (1991) Hydrographic conditions in Icelandic waters 1980–1989. *ICES, Variability Symp.*, No. 6, 10 pp.

Malberg, S.Aa. & Stefansson, U. (1969) Recent changes in the water masses of the East Icelandic current. *ICES, Variability Symp.*, No. 3, 5 pp.

Martin, J.H.A. (1969) Marine climatic changes in the North-East Atlantic, 1900–1966. *ICES, C.M. 1969, Symposium on 'Physical Variability in the North Atlantic'*, No. 23, 15 pp.

Martin, W.R. & Kohler, A.C. (1965) Variation in recruitment of cod (*Gadus morhua* L.) in southern ICNAF waters, as related to environmental changes. *ICNAF Spec. Publ.*, **6**, 833–46.

Marumo, R. (1957) Plankton as the indicator of water masses and ocean currents. *Oceanogr. Mag.*, **9**(1), 55–63.

Matthaus, W. (1991) Long-term trends and variations in hydrographic parameters during the present stagnation period in the central Baltic deep water. *ICES, Variability Symp.*, No. 38, 17 pp.

Meade, R.H. & Emery, K.D. (1971) Sea level as affected by river runoff, eastern United

States. *Science*, **173**, 425–8.

Mehl, S. (1989) The Northeast Arctic cod stocks consumption of commercially exploited prey species in 1984–1986. *Rapp. P.-V. Réun. Cons. Int. Explor. Mer.*, **188**, 185–205.

Midttun, L., Nakken, O. & Raknes, A. (1981) Variations in the geographical distribution of cod in the Barents Sea in the period 1977–1981. *Fisken Hav.*, 1981(4), 1–16.

Mohr, H. (1964) Changes in the behaviour of fish due to environment and motivation and their influence on fishing. *ICNAF Environmental Symposium, Rome*, Contrib. G-3.

Moksness, E. & Øiestad, V. (1987) Interaction of Norwegian spring-spawning herring larvae (*Clupea harengus*) and Barents Sea capelin larvae (*Mallotus villosus*) in a mesocosm study. *J. Cons. Int. Explor. Mer.*, **44**, 32–42.

Möller, H. (1982) Effect of jellyfish predation on larval herring in Kiel Bight. ICES, C.M. 1982/J:11, 8 pp.

Mörner, N.A. (1984a) Climatic changes on a yearly to millenial basis – An introduction. In *Climatic Changes on a Yearly to Millenial Basis* (Ed by N.A. Mörner & W. Karlen), pp. 1–30. D. Reidel Publishing Co., Dordrecht.

Mörner, N.A. (1984b) Concluding remarks. In *Climatic Changes on a Yearly to Millenial Basis* (Ed. by N.A. Mörner & W. Karlen), pp. 637–651. D. Reidel Publishing Co., Dordrecht.

Munk, P. & Kiorboe, T. (1984) First feeding of larval herring at low food concentrations. ICES, C.M. 1984/L:15, 7 pp.

Murata, M., Ishil, M. & Araya, H. (1976) The distribution of the oceanic squids, *Ommastrephes bartrami* (Lesuer), *Onychoteuthis boreali japonicus* Okada, *Gonatopsis borealis* Sasaki and *Todarodes pacificus* Steenstrup in the Pacific Ocean off northeastern Japan. *Bull. Hokkaido Reg. Fish. Res. Lab.*, **41**, 1–30.

Nakken, O. & Raknes, O. (1984) On the geographical distribution of Arctic cod in relation to the distribution of bottom temperatures in the Barents Sea, 1977–1984. ICES, C.M. 1984/GM:20, 10 pp.

Nakken, O. & Raknes, O. (1987) The horizontal distribution of Arctic cod in relation to the distribution of bottom temperatures in the Barents Sea, 1978–1984. Cod Symp., Seattle, 1987, Mimeo 14 pp.

Nehring, D. (1991) Inorganic phosphorus and nitrogen compounds as driving forces for eutrophication in semi-enclosed seas. *ICES, Variability Symp.*, No. 44, 13 pp.

Nehring, D. Schulz, S. & Rechlin, D. (1989) Eutrophication and fishery resources in the Baltic. *Rapp. P.-V. Réun. Cons. Int. Explor. Mer.*, **190**, 198–205.

Øiestad, V. (1983) Growth and survival of herring larvae and fry (*Clupea harengus*, L.) exposed to different feeding regimes in experimental ecosystems: outdoor basins and plastic bags. MS Rpt. Inst. of Mar. Res., Bergen, 299 pp.

Øiestad, V., Kvenseth, P.G. & Pedersen, T. (1984) Mass-production of cod fry (*Gadus morhua* L.) in a large basin in Western Norway – a new approach. ICES, C.M. 1984/F:16, 9 pp.

Ojaveer, E. (1989) Population structure of pelagic fishes in the Baltic. *Rapp. P.-V. Réun. Cons. Int. Explor. Mer.*, **190**, 17–21.

Olsen, S. & Laevastu, T. (1983) Factors affecting catch of long lines, evaluated with a simulation model of longline fishing. NW and Alaska Fish. Center, Seattle, Proc. Rpt. 83-04, 50 pp.

Overholtz, W.J. & Tyler, A.V. (1985) Long-term responses of the demersal fish assemblages of Georges Bank. *Fish. Bull.*, **83**(4), 507–20.

Palmén, E. (1930) Untersuchungen über die Strömungen in den Finland umgebenden Meeren. *Soc. Sci. Fenn. Comm. Phys. Math.*, **5**, 12 pp.

Palmen, E. & Newton, C.W. (1969) Atmospheric circulation systems. Their structure and physical interpretation. *International Geophysics Series*, Vol. 13. Academic Press, New York & London, 603 pp.

Pálsson, O.K. (1984) Studies on recruitment of cod and haddock in Icelandic waters. ICES, C.M. 1984/G:6, 18 pp.

Parsons, T. & Miyake, M. (1989) Biological-physical oceanographic climate study in the subarctic Pacific Ocean. Mimeo, 7 pp.

Radovich, J. (1976) Catch-per-unit-of-effort: Fact, fiction, or dogma. *Calif. Coop. Ocean Fish. Invest. Rep.*, **18**, 31–3.

Radovich, J. (1982) The collapse of the California sardine fishery. What have we learned? *CalCOFI Rep.* **23**, 56–78.

Ratcliffe, R.A.S. (1970) Sea temperature anomalies and long-range forecasting. *Quart. J. Roy. Met. Soc.*, **96**(408), 337–8.

Ratcliffe, R.A.S. & Murray, R. (1970) New lag associations between North Atlantic sea temperature and European pressure applied to long-range weather forecasting. *Quart. J. Roy. Met. Soc.*, **96**(408), 226–46.

Riemann, B. & Hoffmann, E. (1991) Ecological consequences of dredging and bottom trawling in the Limfjord, Denmark. *Mar. Ecol. Progress Ser.*, **69**, 171–8.

Robinson, M.K. (1960) Statistical evidence indicating no long-term climatic change in the deep waters of the North and South Pacific Oceans. *J. Geophys. Res.*, **65**(7), 2097–115.

Robinson, M.K. (1976) Atlas of North Pacific Ocean monthly mean temperatures and mean salinities of the surface layer. U.S. NavOcean Ref. Pub. 2.

Robinson, M.K., Bauer, R.A. & Schroeder, E.H. (1979) Atlas of North Atlantic-Indian Ocean monthly mean temperatures and mean salinities of the surface layer. U.S. NavOcean Ref. Publ. 18.

Rodewald, M. (1966) Abkühlungstrend im Oberflächenwasser des Nordatlantischen Ozeans. *Umschau*, **23**, 777.

Rodewald, M. (1971) Temperature conditions of the North and Northwest Atlantic during the decade 1961–1970. *ICNAF Env. Symp.*, 34 pp.

Rogalla, E.H. & Sahrhage, D. (1960) Heringsvorkommen und Wassertemperatur. *Fischw.*, **7**(5–6), 135–8.

Roll, H.U. (1965) *Physics of the Marine Atmosphere*. Academic Press, New York, 426 pp.

Rollefsen, G. (1932) Fortsatte undersøkelser øver torskegget. *Årsbo. Norg. Fisk.*, 1931(2), 92–7.

Rose, G.A. & Leggett, W.C. (1988) Atmosphere-ocean coupling in the northern Gulf of St Lawrence: Frequency-dependent wind-forced variations in nearshore sea temperatures and currents. *Can. J. Fish. Aquat. Sci.*, **45**, 1222–33.

Rosenfield, A. (1976) Infectious diseases in commercial shellfish on the middle Atlantic coast. *Am. Soc. Limnol. Oceanogr., Spec. Symp.*, **2**, 414–23.

Rossov, V.V. & Kislyakov, A.G. (1969) The polar front in the North Atlantic (according to data collected by R/V *Atlantida*) ICES, C.M. 1969, No. 29, 14 pp.

Russell, F.S., Southward, A.J., Boalch, G.T. & Bitler, E.T. (1971) Changes in biological conditions in the English Channel off Plymouth during the last half century. *Nature* **234**, 468–70.

Saelen, O.H. (1963) Studies in the Norwegian Atlantic Current. Part 2. Investigations during the years 1954–1959 in an area west of Stad. *Geofys. Publ.*, **23**(6), 82 pp.

Saetersdal, G. & Loeng, H. (1983) Ecological adaptation of reproduction in Arctic cod. *PINRO/INR Symp. on Arctic cod.* St Petersburg, Sept. 1983, 23 pp.

Sawyer, J.S. (1965) Notes on the possible physical causes of long-term weather anomalies. *WMO Tech. Note*, No. 66: 227–48.

Scarnecchia, D.L. (1984) Climatic and oceanic variations affecting yield of Icelandic stocks of Atlantic salmon (*Salmo salar*). *Can. J. Fish. Aquat. Sci.*, **41**, 917–35.

Schneider, D.C. & Methven, D.A. (1988) Response of capelin to wind-induced thermal events in the southern Labrador Current. *J. Mar. Res.*, **46**, 105–18.

Scholes, P. (1982) The effect of wind direction on trawl catches: an analysis of haul-by-haul data. *J. Cons. Int. Explor. Mer.*, **40**(1), 81–93.

Schott, G. (1935) *Geographie des Indischen und Stillen Ozeans*. Boysen, Hamburg, 413 pp plus 37 maps.

Schott, G. (1944) *Geographie des Atlantischen Ozeans*. Boysen, Hamburg, 438 pp. plus 27 maps.

Seckel, G.R. (1972) Hawaiian-caught skipjack tuna and their physical environment. *Fishery Bull.*, **70**(3), 763–87.

Seckel, G.R. & Waldron, K.D. (1960) Oceanography and the Hawaiian skipjack fishery. *Pacific Fisherman*, **58**(3), 11–13.

Shannon, L.V., Crawford, R.J.M. & Duffy, D.C. (1984). Pelagic fisheries and warm events: A comparative study. *South African J. Sci.*, **80**, 51–60.

Simrad (1964) *Fish-Finding with Sonar*. Simrad, Oslo, 96 pp.

Simpson, A.C. (1953) Some observations on the mortality of fish and distribution of plankton in the southern N. Sea during the cold winter 1946–7. *J. Cons. Int. Explor. Mer.*, **19**, 150–77.

Sjöblom, V. (1978) The effect of climatic variations on fishing and fish populations. *Fennia*, **150**, 33–7.

Skud, B.E. (1981) Interactions between pelagic fishes and the relation of dominance to environmental conditions. ICES, C.M. 1981/H:60, 11 pp.

Skud, B.E. (1982) Dominance in fishes: The relation between environmental factors and abundance. *Science*, **216**, 144–9.

Skud, B.E. (1983) Interactions of pelagic fishes and the relationship between environmental factors and abundance. *FAO Fisheries Report*, **291**(3), 1133–40.

Smed, J. (1980) Temperature of the waters off southwest and south Greenland during the ICES/ICNAF salmon tagging experiment in 1972. *Rapp. P.-V. Réun. Cons. Int. Explor. Mer.*, **176**, 18–21.

Smed, J. (1983) History of International North Sea Research (ICES). In *North Sea Dynamics* (Ed. by J. Sündermann & W. Lenz), Springer-Verlag, Berlin, Heidelberg, 25 pp.

Smed, J., Meincke, J. & Ellett, D.J. (1983) Time series of oceanographic measurements in the ICES area. Time series of ocean measurements. In *World Climate Programme Report* WCRP, pp. 225–44. World Met. Org., Geneva.

Smith, P.E. (1978) Biological effects of ocean variability: Time and space scales of biological response. *Rapp. P.-V. Réun. Cons. Int. Explor. Mer.*, **173**, 117–27.

Southward, A.J. & Boalch, G.T. (1986) Aspects of long term changes in the ecosystem of the western English Channel in relation to fish populations. *Int. Smyp. Long Term Changes Mar. Fish Pop.*, pp. 415–47. Vigo 1986.

Southward A.J., Boalch, G.T. & Maddock, L. (1988) Fluctuations in the herring and pilchard fisheries of Devon and Cornwall linked to change in climate since the 16th century. *J. Mar. Biol. Ass. U.K.*, **68**, 423–45.

Stefánsson, U. (1969) Near-shore fluctuations of the frontal zone southeast of Iceland. ICES, C.M. 1969, No. 2, 5 pp.

Stommel, H. (1954) Serial observations of drift current in the central North Atlantic ocean. *Tellus*, **6**(3), 203–14.

Sugimoto, T., Kobayashi, K., Matsushita, K., Kimura, K. & Choo, H.S. (1991) Circulation and transport environment for sardine eggs and larvae in Tosa bay. In *Long-term Variability of Pelagic Fish Populations and their Environment* (Ed. by T. Kawasaki). Pergamon Press, pp. 91–9. Oxford.

Sundby, S. & Sunnanå, K. (1990) The North-East Arctic Cod. Appendix III of ICES C.M. 1990/G:50, Report of the Study Group on Cod Stock Fluctuations, pp. 91–138.

Sunnanå, K. (1984) Stomach contents of cod, haddock and saithe on the More coast in 1982 and 1983. ICES, C.M. 1984/G:56, 16 pp.

Sveinbjörnsson, S., Astthorsson, O.S. & Malmberg, S.A. (1984) Blue whiting feeding migration in relation to environmental conditions in the area between Iceland and Faroes in June 1983. ICES, C.M. 1984/H:24, 8 pp.

Svendsen, E. & Magnusson, A.K. (1991) Climatic variability in the North Sea. *ICES, Variability Symp.*, No. 10, 18 pp.

Tabata, S. (1975) The general circulation of the Pacific Ocean and a brief account of the oceanographic structure of the North Pacific Ocean. Part I — Circulation and Volume Transports. *Atmosphere*, **13**(4), 133–68.

Tabata, S. (1983a) Oceanographic factors influencing the distribution and migration of salmonid in the northeast Pacific Ocean. Inst. of Ocean Sciences, Sidney, B.C., Mimeo, 57 pp.

Tabata, S. (1983b) Interannual variability in the abiotic environment of the Bering Sea and the Gulf of Alaska. In *From Year to Year* (Ed. by Warren S. Wooster), pp. 139–45. Washington Sea Grant Publication.

Tait, J.B. (1953) Long-term trends and changes in the hydrography of the Faroe-Shetland Channel region. *Papers in Marine Biology and Oceanography*, pp. 482–98. Pergamon Press Ltd., London.

Tanaka, S. (1983) Variation of pelagic fish stocks in waters around Japan. *FAO Fisheries Report*, **291**(2), 17–36.

Tereshenko, V.V. (1980) The effect of hydrometeorological factors on the transfer of eggs and larvae of the North-eastern Arctic cod into the Bear Island-Spitsbergen area. ICES, C.M. 1980/L:12, 12 pp.

Tiews, K. (1983) Über die Veränderungen im Auftreten von Fischen und Krebsen im Beifang der deutschen Garnelenfischerei während der Jahre 1954–1981. Ein Beitrag zur Ökologie des deutschen Wattenmeeres und zum biologischen Monitoring von Ökosystemen im Meer. *Arch. Fisch. Wiss.*, **34**, Beih. 1, 1–156.

Tokioka, T. (1983) Influence of the ocean on the atmospheric global circulations and short-range climatic fluctuations. *FAO Fisheries Report*, **291**(3), 557–86.

Tormosova, I.D. (1983) Variation in the age at maturity of the North Sea haddock, (*Melanogrammus aeglefinus*) (Gadidae). *J. Ichthyol.*, **3**, 68–74.

U.S. Weather Bureau (1959) Circular R: Preparation and use of weather maps at sea, a guide for mariners. U.S. Dept. of Commerce, 121 pp.

Verploegh, G. (1967) Observation and analysis of the surface wind over the ocean. *Kon. Nederl. Meteorol. Inst. Mededel. en Verh.*, **89**, 68 pp.

Vilhjalmsson, H. (1983) On the biology and changes in exploitation and abundance of the Icelandic capelin. *FAO Fisheries Report*, **291**(2), 537–53.

Vilhjalmsson, H. (1987) Acoustic abundance estimates of the capelin in the Iceland-Greenland-Jan Mayen area in 1978–1987. *Int. Symp. On Fish. Acoustics*, 1987, 22 pp.

Villegas, M.L. & Lopez-Areta, J.M. (1986) Annual and seasonal changes in the captures of *Sardina pilchardus* (W. 1792), *Trachurus trachurus* (L. 1758), *Engraulis encrasicolus* (L. 1758) and *Scomber scombrus* (L. 1758) on the coast of Asturias (1952–1985). *Int. Symp. Long Term Changes Mar. Fish Pop.*, pp. 301–319. Vigo 1986.

Walden, H. (1959) Versuch einer statistischen Untersuchung über die Eigenschaften der Windsee bei abnehmendem, Wind. *Dtsch. hydrogr. Z.*, **12**(4), 141–52.

Walden, H. & Schubert, K. (1965) Untersuchungen über die Beziehungen zwischen Wind und Herings – Fangertrag in der Nordsee. *Ber. Dt. Wiss. Komm. Meeresforsch.*, **17**(2), 194–221.

Ware, D.M. (1986) Climate, predators and prey: Behaviour of a linked oscillating system. Mimeo, 13 pp.

Waterman, J.J. & Cutting, C.L. (1960) Weather and the fishing industry. *The Marine Observer*, **30**, 85–90.

Witting, R. (1909) Kenntnis den vom Winde erzeugten Oberflächenstromes. *Ann. Hydrogr. Marit. Met.*, **73**, 193 pp.

WMO, CMM (World Meteorological Organization, Commission of Marine Meteorology) (1960) Relations between maritime meteorology and fishery biology. CMM-III/Dec. 9 (14. VII.1960. Item 22.

WMO (1966) The preparation and use of weather maps by mariners. Tech. Note No. 72, WMO No. 179, TP. 89.

WMO/ICSU (World Meteorological Organization/International Council of Scientific Unions) (1991) An intercomparison of the climates simulated by 14 atmospheric general circulation models.

WMO, WMP/TD No. 425, 35 pp. plus figures.

Wolfe, D.A. (1985) Fossil fuels: transportation and marine pollution. Wastes in the Ocean 4: In *Energy Wastes in the Ocean* (Ed. by I.W. Duedall, D.R. Kester, P.H. Park & B.H. Ketchum), pp. 45–93. John Wiley & Sons, Inc.

Index